From Analogue to Digital Radio

JP Devlin

From Analogue to Digital Radio

Competition and Cooperation in the UK Radio
Industry

JP Devlin
London, UK

ISBN 978-3-030-06579-9   ISBN 978-3-319-93070-1 (eBook)
https://doi.org/10.1007/978-3-319-93070-1

This Palgrave Macmillan imprint is published by the registered company Springer Nature Switzerland AG
The registered company address is: Gewerbestrasse 11, 6330 Cham, Switzerland

Acknowledgements

I would like to express my gratitude to all those who have helped me in the course of researching and writing this book. I would particularly like to thank colleagues at Bournemouth University's Centre for Media History; Emeritus Professor Sean Street whose knowledge of radio history, the radio industry and programme making quite rightly places him as one of the leading academics and practitioners in the field; Professor Hugh Chignell whose academic input in the field of radio history is well documented and whose theoretical approach to the study of radio helped me appreciate the value of radio as a subject for serious academic scrutiny; and Dr Kristin Skoog who provided the necessary encouragement and support as well as proffering considered, thought-provoking and much valued feedback. I would also like to thank Professor Paul Long at the Birmingham School of Media at Birmingham City University for his careful consideration of my research in this area. I would like to thank the archivists at the BBC Written Archives Centre at Caversham and in particular Jess Hogg who offered great assistance and patience at all times. Finally, I would like to thank all those whom I have spoken to or merely read in the course of this research and those who share a passion for radio and the field of radio studies.

In the footnotes, material from the BBC Written Archives Centre is clearly marked BBC WAC (Written Archives, Caversham) followed by the relevant file number. Also, *Ariel* refers to the name of the BBC's own internal publication for staff which is useful for scrutinizing policy which the corporation felt necessary to disseminate among its employees.

CONTENTS

CHAPTER 1

Introduction

During the course of 2002 the British Broadcasting Corporation (BBC) launched five new national radio stations.[1] This was the first time a new national BBC network had been established since BBC Radio 5 went on-air in August 1990[2] and before that the creation of Radios 1, 2, 3 and 4 back in 1967. Outlining her plans for the five new national radio services on 28 September 2000, Jenny Abramsky (Director, BBC Radio)[3] emphasised the fact that the services would:

> … aim to attract audience groups under-served by existing BBC Network Radio, such as young families and people from ethnic minority backgrounds. Not only would the commissioning of new, quality programming be paramount, the new stations would also draw on the BBC's speech and music archives and give wider exposure to events, such as sports, where the corporation already owned the rights.[4]

Also, simply extending the range of BBC Radio output by creating distinctive, public service radio stations, Abramsky highlighted another

---

[1] BBC Radio 5 Live Sports Extra launched on 2 February, Radio 6 Music on 11 March, Radio 1Xtra on 16 August, Asian Network on 28 October and Radio 7 on 15 December.

[2] Relaunched as BBC Radio 5 Live in March 1994.

[3] Abramsky was appointed Director of Radio in 1999. Her job title was changed to Director of Audio and Music in 2006 and she retired from this position in 2008.

[4] BBC Press Release. 28 September 2000.

© The Author(s) 2018
JP Devlin, *From Analogue to Digital Radio*,
https://doi.org/10.1007/978-3-319-93070-1_1

priority, namely, the need to take advantage of emerging digital platforms, and emphasised that:

> ... each of the new services would be available via digital satellite, digital cable, the internet and digital radio sets.[5]

The unique feature of the new services was that they would be digital services and therefore not available through the ubiquitous, standard, analogue radio receiver which had dominated the radio listening market for the previous 75 years.

Changes to its radio portfolio had already occurred on a few occasions throughout the corporation's history. The BBC had fostered a strong relationship with its audience during the 1920s and 1930s but had also encountered a formidable foe in the form of unlicensed commercial broadcasters bombarding British listeners with popular programming from abroad. With the Second World War looming, the corporation became more alert to the varying needs of its audience and began to consider the necessity of catering for differing listener tastes and aspirations. This saw the eventual creation of the Empire Service and the emergence of the Home Service from the National and Regional Programmes. By 1946 the then Director General, Sir William Haley, recognised a need to reflect a wider range of tastes and develop a more settled system of programmes which resulted in the post-war triumvirate of the Home Service, the Light Programme and the Third Programme.

Commentators often describe the 1930s as the 'Golden Age of Wireless',[6] but in many ways the post-war years represented a zenith of radio broadcasting in the UK,[7] fulfilling a significant role as informer, educator and entertainer[8] during these austere years, even as the new medium of television began to encroach on radio's erstwhile hegemonic status. Although BBC Radio faced little radio competition at this time, the steadily increasing demand for television and the arrival of Independent

---

[5] ibid.

[6] Briggs uses this title for Volume Two of his *History of Broadcasting in the United Kingdom* (Briggs 1995).

[7] Many landmark programmes were created during this period such as *Under Milk Wood* (1954), *The Radio Ballads* (1958), *Hancock's Half Hour* (1954), *The Goons* (1951) and *The Archers* (1951).

[8] First Director General John Reith's description of the role of the BBC to inform, educate and entertain is still at the core of the organisation today.

Television (ITV) in 1955[9] meant BBC Radio was to suffer, as the medium in ascendancy was favoured in terms of resource allocation—particularly after the success of television during coverage of the coronation of Queen Elizabeth II on 2 June 1953, marking it as a 'watershed in broadcasting history' (Holmes 2005, 47).

Since Haley's changes in 1946, the face of BBC Radio remained unaltered for a further 21 years. It was competition from offshore pirate stations which would send a shot over the prow on Portland Place.[10] The BBC could not ignore the popularity of the popular music service provided by the pirates during the 1960s to an avid young audience, which the corporation had been neglecting, and Radio 1 was deemed the perfect vehicle for this disaffected group to board.[11] As the pirates were eventually banned under the Marine, &c., Broadcasting (Offences) Act 1967 and Radio 1 came to air; the Light Programme, the Third Programme and the Home Service were respectfully renamed Radios 2, 3 and 4.

In 1973, BBC Radio had to emerge from its cocoon as LBC, and Capital Radio in London became the first legalised commercial contenders in the marketplace. Such competition was to spread on a local level and eventually on a national level as Classic FM became the first national offspring of the 1990 Broadcasting Act.[12] As more and more stations entered the arena, offering alternative listening brands for an increasingly demanding and selective audience, the weaknesses of the BBC national networks became exposed through a plummet in listening figures. For example, Radio 2's share in London dropped from 22% in 1988 to 11% in 1991,[13] and by 1993 Radio 1 had lost around eight million listeners to commercial alternatives.[14] By the time of the launch of the new digital services in 2002 however, BBC national radio had undergone something of a renaissance and was again beginning to enjoy successful listening across all networks with Radio 4 pushing 10 million listeners and Radio 2

---

[9] BBC Radio went head to head on ITV's opening night with the death of Grace Archer in *The Archers*.

[10] Architect George Val Myer used the idea of an ocean liner as the design for BBC Broadcasting House in London. See Reid (1987).

[11] DJ Tony Blackburn launched BBC Radio 1 on 30 September 1967.

[12] Classic FM launched in September 1992 followed by Virgin 1215 in April 1993 and Talk Radio in February 1995.

[13] BBC Radio Daily Survey Reports 1991.

[14] BBC Radio Daily Survey Reports 1993.

consistently reaching 12.5 million.[15] Radio as a medium had been holding its own against other media with RAJAR[16] figures suggesting that for the first time in at least three decades, radio listening, both BBC and commercial, had exceeded television viewing in the UK.[17] Street (2002, 135) proposes that history will judge May 2002 as a 'significant milestone in radio history' as audience figures began to signal a recovery of the BBC's position and also demonstrate a strong performance for the commercial sector and hence the industry as a whole.

This cursory summation of the history of the UK radio industry in the twentieth century is intended to briefly outline three aspects of radio history which this book aims to examine—the BBC, the commercial radio sector and the relationship between the two. It attempts to account for the long-term success of the medium and the roles played by the BBC and the commercial sector in ensuring that longevity. It is not a series of linear histories covering well-trodden ground, but instead represents a historiography which examines each in relation to the other, identifying the particular roles played by each in maintaining the continued survival of the industry as a whole. I intend to develop an argument which proposes the BBC and commercial radio played distinct and separate roles during the era of analogue radio, with very little common direction, but that the digital era heralded a new and unique period of cooperation and that this working in concert shifted the established dynamic within the UK radio industry. So, with the move in technology from analogue to digital there also came a change in relationships—from competition to cooperation. The history of radio in the UK is typified by singular actions by the public service broadcaster and its commercial opponents in various guises. Such actions have been important in helping radio maintain its powerful media presence and I wish to highlight the crucial roles each party has played. I also seek to underscore the joint actions taken in order to launch digital audio broadcasting (DAB) digital radio and demonstrate how this model of cooperation was both novel and indeed fruitful in its simple task of keeping radio at the forefront of an ever-expanding media environment.

A mere historical account does in itself make fascinating reading for any student of radio, but beyond the timeline approach it is also critical on an academic level to illustrate this history of radio within a comparative

[15] RAJAR Report, 2003. London: RAJAR, 2003.

[16] Radio Joint Audience Research Limited) was established in 1992 to operate a single audience measurement system for the UK radio industry.

[17] RAJAR Report, 2001. London: RAJAR, 2001.

context, namely as a succession of relationships between the BBC and the commercial radio sector. If narratives of radio history tend to be skewed heavily towards the BBC, then this book aims to demonstrate how both the BBC and the commercial sector have played important roles in radio's position in the media terrain while at the same time illustrating how the move from analogue to digital radio represented a landmark in the history of radio in the UK—one defined not just by the introduction of a new technology but also in the shifting of relations between the industry's two leading players. Also, by highlighting the steps taken by each party to bolster radio's general success, I seek to demonstrate how a previously distant rapport was transformed—for a period—into one of overt cooperation in order to promote a technology which both parties envisaged as necessary for the survival of the greater industry as a whole. This is a key period in the history of the UK radio industry but also in the history of BBC Radio and in the history of UK commercial radio, when the dynamic of existing relations altered. By questioning why such a change took place and illustrating its form, the study provides an insight into a unique era and one which represents a significant deviation from the preceding eras.

There are two broad questions which this account seeks to address. Firstly, what has been the historical relationship between the BBC and the commercial sector, and secondly, how did that relationship change in order to promote DAB digital radio? In answering these questions it will be possible to ascertain how different parties have been responsible for radio's success at different times during the course of its history, and largely in an atmosphere of unyielding competitiveness, before demonstrating how a significant change of dynamic within the industry led to a productive accomplishment for all parties.

Chapter 2 examines the origins of broadcasting in the UK, how an industry emerged from a technology and how that in turn produced two distinctive models of broadcasting which have persisted to the present day, that is, public service broadcasting and commercial, profit-driven enterprise. Chapter 3 covers the period of the Second World War and the postwar years when competition was eradicated and the BBC remained as sole supplier of content but was also left with time to re-establish its position of primacy and prepare to maintain this position in the relative quiet of the 1950s. The 1960s was to be different, it was to be a period of intense competition for the BBC and one which would only come to an end with the aid of legislation, and this is covered in Chap. 4. Legislation would not just eradicate competition, it would at a later date actually enable it, and

Chap. 5 describes the arrival of legalised, commercial broadcasting in the form of *independent* radio at a local level. Chapter 6 goes on to examine the next stage of licenced broadcasting in the form of fully fledged *commercial* radio and not just at the local level. Finally, on a more level playing field, Chap. 7 considers the arrival of DAB digital radio and how that shaped existing relationships within a now more balanced industry. I hope to illustrate the static and unyielding relationship between the BBC and its commercial rivals through the period of analogue radio before exploring how, after decades of divergence, both the BBC and the commercial sector converged in order to secure the future of the radio industry through the promotion of DAB and thus reversed a long established division within the industry. Over the course of the twentieth century each party has played its own separate role in radio's continued existence and it is important that these roles are marked so as to attribute the necessary recognition. By the end of the century, as radio underwent a huge technological transformation, then an act of joint enterprise ought to be similarly recognised.

It is useful to point out what this book does not cover. I use the term commercial radio throughout as a general umbrella term to cover all commercial competitors, whether legal or illegal or whether broadcasting from abroad or from within the UK. Of course different variations of the term will apply in different eras but all are identifiable as commercial enterprises, that is, business models which aim to make a profit from broadcasting, usually through selling advertising and even when under legislative restrictions. Community radio is an important part of the radio gamut but one which strictly does not fit into the commercial radio category in this study as it has had a smaller—albeit important—role to play and has not been a driver of radio on a scale that has influenced the continued success of the medium at the macro level. I also do not seek to cover it in depth simply because it is not usually in direct competition with the BBC or other radio companies because its audience is so niche and also because, in terms of academic studies, it is an area which is already very well covered. This of course is a study of radio, and there is no doubt television (both BBC and commercial) has had a huge effect on radio. Where necessary, I do examine the impact of television at certain critical points, but to consider the greater role of television and its impact would require a longer, separate inquiry. As this is a study of the UK radio industry, I have avoided using international comparisons. Although in itself this would be an area worthy of scrutiny, I have decided to concentrate on the UK radio model as a centrepiece of this work and leave a comparative analysis for consideration by a future author.

The main body of the study finishes in 2002/2003 and I have chosen this cut-off point for a number of reasons. This was the point where the BBC launched new digital only services, thus demonstrating a long-term commitment to DAB, and it is the point where the commercial sector had firmly established its DAB position. While some elements within the commercial sector can be seen to falter after this date, it represents what I suggest was the pinnacle of the commercial sector's embracing of DAB. This date also represents the end of the period of active cooperation and therefore the end of a unique era in the history of the long relationship between the BBC and commercial radio. I believe 2002/2003 may be recorded as the zenith of not just the arrival of DAB in the UK but also of the relationship between the BBC and its rivals.

Due to the specific nature of the subject under consideration, this study rests firmly within the discipline of 'radio studies' but also contributes to the field of 'media history' which in itself has been referred to as either the 'neglected child of media studies' (Brugger and Kolstrup 2002) or the 'neglected grandparent of media studies' (Curran 2002, 3). The reason for its neglect as a subject for serious consideration—whichever familial position it holds—is sometimes levelled at the scepticism within the academic community itself and particularly among traditional historians who were mostly inclined to be dismissive. Writing in 1994 Dahl (1994) asks:

> What exactly is the object of media history and to what extent is there
> such a thing as a proper historical study of the communication media,
> recognizable as a discipline on its own merits and not just as a
> hyphenated offshoot of technological or cultural history?

In replying to this, one can cite O'Malley (2002) who believes that traditionally there was little concern regarding the historical influence of the media, but after 2000 this changed as the study of the history of individual media forms became more developed as the result of:

> ... a slow realisation by academics that the mass media and communications
> were pervasive elements of nineteenth and twentieth century societies and as
> such had to move from the margins towards the centre of historical
> investigation.

Thus we see in more recent times a greater appreciation of the field of media history resulting from the subject's expansion from simple historical account and into the domain of social and cultural subjectivity.

Curran (2002, 3) claims media historians' labour in the shadows because of their 'often narrowly specialized scholarship', but likewise, one of Curran's main contributions to media studies is his insistence on the centrality of history with his goal:

> ... to advance a tradition of media history that seeks ambitiously to situate historical
> investigation of the media in a wider societal context. (Curran 2009, 20)

Avoiding a narrow historical remit is central to this study. Rather than succumb to any accusation of merely adding to a well-established narrative of the history of radio, I hope to pursue a wider social and organisational history (in terms of the interplay between actors in the radio industry), thus portraying an aspect of media history with wider repercussions.

The inherent danger in any historical study of the British media, particularly in relation to television or radio, is the pre-eminent role played by the BBC and therefore the tendency to produce a single layered institutional history. Briggs (1995) is the seminal history of the BBC—and worthy of that accolade—but has been challenged by some media historians for providing a standardised corporate history. For example, Hajkowski (2010, 8) notes:

> In addition to the organization and development of radio, Briggs
> examines the relationship between the BBC and the government,
> technological change, the manufacture and marketing of receivers,
> and the impact of radio on social habits and leisure. However, Briggs
> ultimately produces a top-down, institutional history of the BBC...
> As a consequence, Briggs's *History* gives short shrift to certain
> aspects of broadcasting.

Hajkowski is concerned that the treatment of interdisciplinary issues is limited with too much emphasis on broad analyses, and Briggs (1980) himself did go on to lament the omission of a 'greater study of cultural factors' in his magnum opus.[18] It is important to avoid the pitfall of solely examining the BBC's role as the dominant player without adding equal weight to its competitors and their social and cultural influences and effects. While any study of the UK radio industry will necessarily involve a significant appraisal of the BBC due to its position, both historical and

---

[18] Hajkowski levels the same criticism at another influential work of broadcasting history; Scannell and Cardiff (1991).

contemporary, I wish to obviate the tendency in such studies to attach a pre-eminent role for the BBC and indeed challenge any such assertion. Rather than representing a mere adjunct to BBC radio history, this book instead demonstrates how shifts in the BBC's position came about and placed it on a more equal footing with the commercial sector. As for the commercial sector, again I avoid the simple historical narrative and like-wise focus on its rapport with the BBC, and in doing so reveal an often overlooked attribute of the former, that is, its equally crucial role in the history of radio, none more so than in the DAB era, thus adopting an often neglected position, namely, one that celebrates the role played by commercial radio. In historiographical terms, this avoids the danger often associated with media studies in general which Pickering (2015, 16) has identified as proceeding:

> … without an integral historical perspective informing its key questions
> and pre-occupations, and without an active historicizing impulse … This is
> precisely why, against the strident emphasis on newness and nowness,
> we need to be … more receptive to slower processes of cultural change
> and adaptation, longer-term institutional formations and resilient structural
> continuities.

A desire to pursue research in this area emanates from a very simplistic personal observation, which was the sudden appearance of competing and distant forces conjoining on platforms to herald a new era of camaraderie. As a BBC member of staff since the early 1990s, I had been acutely aware of the distance between BBC Radio and its competitors, but by the end of the 1990s this had changed. Suddenly I was attending joint meetings and conferences and indeed pub evenings where the parties gathered to share in a common mission to promote DAB. From this, questions arose as to what were the driving forces behind this and what necessitated it? What was then required was an objective study into the exact historical nature of the relationship in order to accurately define its status, before proceeding to ascertain how this status changed and why. In studying the history of relations between the BBC and the commercial radio sector and then demonstrating how such relations were altered by the emergence of a new technology, I hope to address a lacuna in radio history by highlight-ing the significant roles played by both the commercial sector and the BBC over the course of the twentieth century in order to ensure the con-tinued success of radio and how a policy of unprecedented cooperation would continue to fulfil that role. In doing so, I hope to add to the continued academic study of radio itself and hope this contribution meets the ongoing challenge of helping radio emerge from what Garner (2003) describes as 'the shadows of critical neglect.'

References

Briggs, A. "Problems and Possibilities in the Writing of Broadcast Histories." *Media, Culture and Society* 2, no. 1 (1980).

Briggs, A., *The History of Broadcasting in the United Kingdom*. Vols 1–5. Oxford: Oxford University Press, 1995.

Brugger, N. and Kolstrup, S. *Media History: Theories, Methods, Analysis*. Aarhus: Aarhus University Press, 2002.

Curran, J. *Media and Power*. London: Routledge, 2002.

Curran, J. "Narratives of Media History Revisited." In *Narrating Media History*, edited by Bailey, M. Abingdon: Routledge, 2009.

Dahl, H. "The Pursuit of Media History." *Media, Culture and Society* 16, no. 4 (1994).

Hajkowski, T. *The BBC and National Identity in Britain, 1922–53*. Manchester: Manchester University Press, 2010.

Garner, K. "On Defining the Field." *The Radio Journal: International Studies in Broadcast and Audio Media* 1, no 1 (2003).

Holmes, S. *British Television and Film Culture in the 1950s*. Bristol: Intellect Books, 2005.

O'Malley, T. "Media History and Media Studies: Aspects of the Development of the Study of Media History in the UK, 1945–2000." *Media History* 8, no. 2 (2002).

Pickering, M. "The Devaluation of History in Media Studies." In *The Routledge Companion to British Media History*, edited by Conboy, M and Steel, J. Abingdon: Routledge, 2015.

Reid, C. *Action Stations: A History of Broadcasting House*. London: Robson Books, 1987.

Scannell, P. and Cardiff, D. *A Social History of British Broadcasting 1922–1939*. Oxford: Blackwell, 1991.

Street, S. *A Concise History of British Radio 1922–2002*. Tiverton: Kelly Publications, 2002.

CHAPTER 2

From New Technology to New Industry: The Emergence of Radio Broadcasting in the UK

Radio, throughout most of its history, has been readily definable. It is a method of broadcasting, that is, the transmission of words and music over radio waves, distributed by a provider and picked up by a listener on a receiving set, where the listener can enjoy the content provided or, if not, simply switch to another provider. This model has been familiar to most of us, as consumers of radio broadcasting over the course of the twentieth century. Now, in the second decade of the twenty-first century, we still employ the term radio but it has become much more generic by including formats such as internet streaming, podcasting and audio on demand which differ from the traditional twentieth century model. This traditional model itself took time to become established after the technology had arrived, and various actors played a role in creating it.

It was commercial enterprise which drove radio at its technical development stage, although with an uncertain idea as to its ultimate use, as these commercial interests were not broadcasters but radio receiving set manufacturers. The formation of the British Broadcasting Company represents an act of joint enterprise by these commercial actors which eradicated competition and created the environment for technology to become inextricably linked to content, as the specific role of radio—in broadcasting terms—became established. Clearly, this period was initially driven by technology, but as the focus on radio's position sharpened—and with the influence of legislation—the early flowerings of what we came to perceive

© The Author(s) 2018
JP Devlin, *From Analogue to Digital Radio*,
https://doi.org/10.1007/978-3-319-93070-1_2

as radio in the twentieth century became apparent. This period is characterised by a flurry of competitive activity among radio set manufacturers
before a cooperative endeavour among them to establish a radio industry
based on content, followed by the establishment of a public service model
which would endure for decades. Commercial interest steered radio technology towards a new broadcasting industry which would be consolidated
by a public service model in the shape of the British Broadcasting
Corporation. This period is characterised by competition among set
manufacturers who in turn found it necessary to cooperate in order to
establish radio's actual role beyond its mere technology. Once established,
a single model of broadcasting was deemed essential for an industry to
blossom.

The Pre-broadcasting Era

Guglielmo Marconi arrived in Britain in 1896 to continue his experiments
into 'transmitting electrical impulses and signals' (Marconi 1896) which
he had been carrying out in his native Italy over the previous few years.
Marconi's work followed on from successes in the field from scientists
such as James Clerk Maxwell (Electromagnetic Theory) and Heinrich
Hertz (Electromagnetic Waves), culminating in his groundbreaking, first
transatlantic wireless signal between Cornwall and Newfoundland in
December 1901.[1] The interesting aspect of radio history during these very
early days, at least from our privileged position of hindsight in the mass
communications world of the twenty-first century, is the fact that the
broadcasting dimension of the new technology was not considered in any
way advantageous. In fact, wireless, which represented the culmination of
nineteenth-century communications history, was thought of simply as 'a
mere substitute for wired telegraphy' (Briggs and Burke 2002, 154) and
its use considered to be relevant only in a similar fashion, namely, for military or commercial purposes. Wired telegraphy, the transmission of long-
distance textual or symbolic messages—as opposed to verbal messages—had
garnered much interest in certain sectors. Indeed, on founding his
Wireless, Telegraph and Signal Company in 1897, Marconi himself
regarded customers as being solely large commercial organisations and
governments.

[1] The first transatlantic wireless message from Cornwall, UK, to Newfoundland, Canada,
was received on this date; it was the Morse code 'S' signal.

Wireless telegraphy, which meant the exchange of such messages over the airwaves, would soon demonstrate its potential as an important tool of point to point communication over the early years of the twentieth century and in particular in that very crucial domain of telegraphy: ship to shore. The words 'Good morning, Dr Crippen. Do you know me? I'm Chief Inspector Dew from Scotland Yard' (Young 1920) and the subsequent handcuffing of Dr Hawley Crippen on board the transatlantic steamer SS *Montrose* in 1910 represented the first arrest made with the aid of wireless technology. Also, we witness the key role played by wireless equipment in saving lives on board the *Titanic* in 1912 as two Marconi Company operators in the ship's wireless room sent out a flurry of distress signals soon after the vessel's fateful collision with an iceberg. These are just two from a number of illustrations of the success of this nascent technology and both demonstrated the technology's natural, yet still narrow, aptitudes.

By 1914 most British wireless experts, including leading figures in the Wireless Society of London,[2] were still unconvinced that wireless telegraphy had a future any wider than its existing function of point to point messaging. In the USA, things had been progressing very differently. As early as 1906, R.A. Fessenden[3] had experimented in telephony—the transmission of the human voice, as opposed to signals—by transmitting speech and gramophone records to ships at sea and asking those listening to contact him to confirm if they had heard 'the voices across the air' (Burns 2004, 321) on their receiving sets. Although technically Fessenden was testing traditional ship to shore capabilities, he used a different method of doing so and consequently can be considered one of the first people to employ the technique of 'broadcasting.' Two things characterise Fessenden's experiments, and they would shape broadcasting as we know it today but which at this time was still a long way off. Firstly, he employed the human voice and secondly, rather than a point to point communication, Fessenden was transmitting messages and entertainment from one single source to a wide number of recipients, thus marking his experiment as the 'first radio broadcast' (Sarkar et al. 2006, 409). This notion caught on in the USA and was best extolled by David Sarnoff, an employee of the

---

[2] Formed in 1913 as a society for amateur radio operators and is still going today as the Radio Society of Great Britain http://rsgb.org/.

[3] Reginald Aubrey Fessenden was a Canadian born inventor and pioneer in the field of radio technology.

American Marconi Company, and often labelled as 'the prophet of radio' (Lyons 1966, 117) who described what he envisaged as the true potential and future of wireless by predicting the arrival of radio as a household utility and its ability to act as a:

> Radio music box... arranged for several different wavelengths which should be changeable with the throwing of a single switch or the pressing of a single button. (Sarnoff 1968, 31)

The First World War also heralded a maturation of wireless from mere implement of communication to powerful broadcasting tool, albeit under more inauspicious circumstances. Germany had built what was then the world's most highly powered wireless telegraphy station, at Nauen near Berlin, which it used for the purpose of transmitting propaganda. The British government in turn assumed control of the Marconi works in Chelmsford, and all other wireless transmission activities were also commandeered. The new wireless technology was becoming well established as a tool of propaganda as much as a channel of communication on the battlefield. The war began with British soldiers using telegraphs over wired lines but ended with wireless telegraphy and telephony which in itself changed the very nature of warfare.[4] Wireless was however gradually growing a new limb. One to one wireless telephony, although still crucially important, began to be overshadowed by the envisaged potential of 'wireless broadcasting' which we distinguish as the transmission of radio from one to many as opposed to from one to one. In the years immediately after the First World War, Marconi continued his experiments with broadcasts from Ballybunion in Ireland to the USA in 1919 and the construction of a huge transmitter at Chelmsford which was soon 'testing speech and musical broadcasts which were being picked up all across Europe' (Hennessy 2005, 49). In 1920, newspaper magnate Lord Northcliffe decided *The Daily Mail* should sponsor a special broadcast. So on 15 June the celebrated singer Dame Nellie Melba went to the Marconi works and sang to listeners as far away as Spain and Norway. This was the moment that wireless broadcasting really caught the imagination of the British public and the point from which a demand for broadcasting began to take hold in the UK.

---

[4] Some early equipment is on display at the Royal Signals Museum in Dorset.

THE CONSOLIDATION OF BROADCASTING

The Marconi Company continued its experimental broadcasts from Chelmsford. They were however affected by a growing disdain among those in official circles who were concerned that these broadcasts were interfering with that other arm of wireless technology, namely, military communications. Under intense pressure from such quarters, the Post Office finally decided to ban the Chelmsford broadcasts in 1920. The inevitable vacuum created was filled by the radio amateurs, fanatics of wireless who individually and collectively were either actually involved in setting up transmissions or, as receivers (i.e. listeners), were voicing their disapproval at the silence of the airwaves. These amateurs or *hams* had been forging national and international links using Morse and later telephony and continued very small-scale broadcasting after Marconi's activities had been severely curtailed. However, it was the actions of those more interested in actual listening and who had formed wireless groups across the UK, who exercised the greatest influence on the Post Office to allow regular broadcasts to recommence. In 1920 there were 20 wireless societies across the UK (e.g. the Wireless Society of London), but by 1921 this number had grown to 63 (Smith 1974, 67), and they all had a common demand: the reinstatement of a broadcasting service. In December 1921, representatives of all the societies petitioned the Post Office voicing a:

> ... national resentment that public services such as wireless time
> and telephony should be left to our neighbours to provide. (Briggs 1995, 57)

The British public had tasted the delights of broadcasting, had seen its logical position in the tri-media world of the day—along with newspapers and cinema—and simply wanted more of it, and from British sources rather than from sources based abroad. From a very early stage demand for wireless broadcasting is clearly evident. Although it has still not formed the model which we recognise today, what is clear is that people wished to consume it, and as demand increases then there exists a potential for those who can meet that demand. This basic economic model is one which still pervades in the media industry today.

It is certainly true at this stage that as Britain procrastinated the USA 'forged ahead' (Briggs ibid., 95), both on a practical level and an intellectual level. By June 1922 David Sarnoff was describing broadcasting as representing 'a job of entertaining, informing and educating the nation'

(Sarnoff 1968, 30), a maxim more commonly attributed to John Reith.[5] Across the Atlantic, developments such as the formation of the Radio Corporation of America (RCA) and the setting up of stations by the likes of the General Electric Company (GEC) meant that Americans were enjoying not only musical entertainment but also sports commentary, talks and even church services from the comfort of their own homes, and by May 1922 there were 219 registered radio stations across the USA (Street 2002, 24).

In the UK the Post Office decided to react to the rising tide of demand for a wireless service—particularly from the *hams*—by rescinding the ban on the Marconi Company broadcasts in December 1921, and on 13 January 1922 the Postmaster General authorised the Marconi Company to include, within the weekly period of 30 minutes it was already allowed to experiment, a programme of 15 minutes of telephony comprising both speech and music for the benefit of the wireless societies. The new service, known as 2MT, first broadcast from a hut in Writtle, just down the road from Chelmsford, on 14 February 1922 and represents a landmark not only in its huge significance at the time but also due to the fact that it was at this station where the 'nucleus of the "brains trust" of the technical side of British broadcasting could be found' (Briggs 1995, 58). Not only were the engineers at Writtle to be the very people who would eventually lead future technical developments in broadcasting but they also became adept script writers, producers and entertainers,[6] thus combining technical and creative capabilities which would characterise the evolution of the new industry.

By May 1922 the Marconi Company had secured permission from the Post Office to broadcast from its headquarters on The Strand in Central London. The station, called 2LO, operated at greater power than 2MT, but these two were by no means the only stations now operating in the UK, nor was Marconi the only business enterprise involved as the industry expanded. During 1922 we first witness the new phenomenon of competition of the airwaves among various enterprises. In addition to Marconi, Metropolitan Vickers, Western Electric, General Electric, Radio Communication and British Thomson-Houston were the 'Big Six' (Street

[5] John Reith, later Lord Reith. Held positions of BBC General Manager, Managing Director and Director General 1922–1938.
[6] For example, P.P. Eckersley, who was an engineer and broadcaster at 2MT and would later become BBC Chief Engineer (see Eckersley 1941).

2002, 27), the major players in an industry of numerous companies all striving to implant radio across the UK. The British radio market was wide open but this was leading to two distinct problems. Firstly, the airwaves were becoming cluttered as more and more broadcasters came on board, and secondly, the Marconi Company was rapidly establishing a monopoly as the strongest actor within this group, towering above the many smaller enterprises. In essence, by 1922 the UK radio industry was one characterised by a vast number of small operators creating chaos due to the limited availability of wavelengths while at the same time dominated by a formidable large player, a scenario which would characterise the radio landscape some 70 years later.

In order to avoid chaos and congestion on the airwaves, as was apparent in the USA, the Post Office decided to act and announced it would licence a limited number of broadcasting stations across the country to be run by the manufacturers of wireless apparatus, but in granting such a licence, the Post Office was keen to ensure that no monopoly of broadcasting would emerge nor that any service would emanate from outside the UK. Not wanting to waste any time, the wireless manufacturers arranged to meet at the Institute of Electrical Engineers in London on 23 May 1922. The meeting was attended by a number of companies but steered by the 'Big Six.' This meeting was to represent the first in a series of demarches in British broadcasting that would see the emergence of a single, powerful broadcasting service. In this instance, it was one based on cooperation between competing forces in order to promote a joint venture, and only a few days later the manufacturers' consortium declared the name of their venture as the British Broadcasting Company.

THE BRITISH BROADCASTING COMPANY

The new company was formed on 18 October 1922 with over 200 manufacturing firms present and received its licence from the Post Office on 18 January 1923. The company's funding was to come from its own stock, royalties from the sale of receivers and revenue from broadcasting receiving licences. Daily broadcasts from the existing 2LO studios in Marconi House began on 14 November 1922, and local services then sprung up across the country. By 1925 a national network was established, and with the opening of the Daventry transmitter that same year,[7] around 80–85%

[7] The Daventry transmitter opened on 27 July 1925 and was the world's first long wave transmitting station. http://www.bbc.co.uk/programmes/p027c6h8.

of the UK population could now potentially receive wireless broadcasts. At this point in time, with the onset of a broadcasting structure, it is useful to note the appropriation of nomenclature. The term broadcasting had become the standardised term for the transmission of speech and music by radio waves to a wide audience. In effect, it became a verb with a modern definition, somewhat detached from its erstwhile agricultural meaning of scattering seed by hand over a wide area—a meaning which would be totally lost by the end of the century. Also, it is important to note at this point that the terms radio and wireless, for both the activity and the object, began to be used interchangeably from this time and would continue to do so until the term wireless began to lose currency around seven decades later.

A number of significant developments occurred during the era of the British Broadcasting Company which would prove pertinent for the survival of its successor. In terms of personnel, the appointment of John Reith as General Manager on 14 December 1922 was a crucial decision. A year later Reith would become Managing Director and, a number of years later, Director General, of the newly formed British Broadcasting Corporation. Certainly for many, Reith *was* the British Broadcasting Company, such was his impact throughout the organisation. An engineer by profession, Reith easily assumed the mantle of management and indeed found himself on occasion behind the microphone. It was his evangelism regarding the new culture of broadcasting which largely remains his legacy however, as well as his unwavering stance on the principle of public service broadcasting and standards thereof.[8]

While the formation of the British Broadcasting Company signified a new era of cooperation between competing radio entities, it also exposed the medium's position vis-à-vis other media. Reporting of news on the new wireless service came under a severe restraint which was imposed by the newspapers and press agencies intent on protecting their own interests. The *7 o'clock rule* forbade the broadcasting of news before 7.00pm and any news broadcast after this time could only be scripted from the main agencies. However, during the General Strike of 1926 the restrictions of the Newspaper Proprietors' Association were lifted for the duration, and this helped establish wireless as an important means of disseminating news and important information in a rapid way to a vast audience. The *7 o'clock rule* was eventually lifted in 1938, but despite the

---

[8] See Reith (1924) and Reith (1949) for his own experiences of this era.

political difficulties of reporting the news during the strike, the potential for wireless in the news domain was obvious. The General Strike played a role in developing the future character of the BBC since Reith was anxious to retain a measure of independence from government and this it managed to do, albeit with great difficulty, but it 'succeeded against the odds in using the strike to strengthen the position of the BBC' (Briggs 1995, 330).

As broadcasting became more popular among listeners, there was an increase in the sale of receivers and consequently the number of licences,[9] but this in itself produced financial problems for the new company. Many listeners were also wireless enthusiasts wholly capable of constructing their own receiving sets and this they invariably did.[10] In these instances, and they were many, the company was therefore gaining no income from the sale of their own sets nor from a licence fee. Due to the furore over this blatant hole in the funding process, the government decided to set up an inquiry under the chairmanship of Sir Frederick Sykes to examine the problem. The Sykes Committee[11] reported to Parliament in August 1923 and as well as praising the work of the British Broadcasting Company, it recommended a single licence fee which conferred upon the holder the right to listen to the former's service using whatever type of receiving set. This meant that the wireless licence became the single source of finance for the British Broadcasting Company and the legal obligation upon listeners to purchase a licence became abundantly clear.

By 1925, with receiving sets able to pick up stations across Europe and with the public's insatiable demand for wireless programmes rising, it was time to take stock of the state of British broadcasting, identify its position and ultimately consider its future. To this end the government set up another committee, this time under the chairmanship of the Earl of Crawford and Balcarres which reported to Parliament in March 1926.[12] Reith, by this time, was developing his own philosophy of British broadcasting in which he wanted the British Broadcasting Company to become a wholly public institution free from any commercial or political pressures.

---

[9] 2,178,259 BBC licences were sold in 1926 (Licence Figures, BBC Handbook 1939).

[10] In the 1920s it was estimated there had been over 250,000 people in Britain who had made their own homemade crystal sets (Coyer 2007, 16).

[11] The Broadcasting Committee Report 1923 (Sykes Report). Cmd. 1051. London, HMSO.

[12] Report of the Broadcasting Committee 1926 (Crawford Report). Cmd 2599. London HMSO.

The Crawford Committee thought along similar lines and proposed the establishment, through Royal Charter, of a single, independent organisation charged with the task of providing a service in the national interest. It would be interesting to see what form of broadcasting this might produce after an unsettled period of 'continuing struggle and negotiation' (Scannell and Cardiff 1982, 161).

THE BRITISH BROADCASTING CORPORATION

The government accepted most of the Crawford Committee's recommendations and the new British Broadcasting Corporation (BBC) came into being for an initial period of ten years as of 1 January 1927. Under the terms of the charter[13] the new BBC had a guaranteed income from receiving set licences and assumed full editorial independence, though under the guardianship of a Board of Governors appointed by the government. Indeed, the BBC's constitution and its statutory obligations as a publicly funded yet quasi-autonomous institution, namely, its role to inform, educate and entertain; its duty to report the proceedings of Parliament; its requirement to preserve political balance and not editorialise; its requirement to eschew advertising; and its duty to broadcast government messages in a national emergency, have largely remained unchanged since.

The new corporation had been created at an opportune time as enthusiasm for wireless was still at a high among the public and some form of order on programme provision was desperately needed. The BBC therefore embarked on a genuine effort to complete the steps taken before 1927 in securing a single national service covering the entire UK[14] as well as more focussed regional services.[15] Without doubt the BBC's engineers must be recognised for the important role they played in establishing a network of transmitter stations which allowed the entire country to listen and to become unified in listening to the same programming, and also for each area to listen to content that would only be of interest at a local level. This provision helped the BBC secure for itself an almost sacrosanct position among the listening public. The BBC's desire to use its monopoly

---

[13] Charter for the period 1 January 1927 to 31 December 1936. All BBC Charters are available at http://www.bbc.co.uk/bbctrust/governance/regulatory_framework/charter_archive.html

[14] The National Programme began in March 1930.

[15] The Regional Programmes began in March 1930.

position to create a unique broadcasting model was aided on one hand by technological advances (transmitters) and on the other by governmental authority (charter) which:

> ... strongly endorsed John Reith's eloquent re-articulation of the national function of the BBC, resulting in the transformation of an awkwardly coordinated private enterprise into a ground-breaking public corporation. (Hilmes 2012, 49)

Not only was the BBC stamping its name across the UK but also abroad with the launch of the Empire Service in 1932. Also, due to the growth of the medium, the corporation was to leave Savoy Hill and move to new custom-built accommodation at Broadcasting House in the same year. It was during this period that listening to the wireless became a widespread activity and sales of receivers began to reflect this new found status with an estimated 600,000 radio sets being sold annually by 1930, and this figure rising to around 1.26 million in the early 1930s (Scott 2012). Also around this time many entertainers and serious broadcasters were becoming household names[16] in what may be described as the early days of the concept of the 'media celebrity.' Yet, despite its undoubted hold over the listening public and ability to reach across the land, and the fact that its broadcasters were recognised by huge numbers of people, a threat to the BBC's position was emerging. The corporation felt comfortable in its role as a national broadcaster but, during the period of its first charter, the airwaves began to fill with other voices from other sources providing engaging programming. Radio broadcasting had become a commodity and one which other actors thought they could also provide and, more importantly, derive profit from.

The British Broadcasting Company had eradicated open competition in the UK radio arena, but it had done so by instilling an ideal of cooperation between competing entities for a common good. The formation of the BBC completely changed the radio landscape. There would be no competing forces operating from within the new organisation. Instead it would be a single body responsible for all broadcasting in the UK, one which would be responsible for broadcasting on both a regional and national level. It would be the ultimate broadcasting monopoly, a single entity

---

[16] Names such as A.J. Alan, Arthur Burrows, Emma Barcroft, Sydney Firman, Carroll Gibbons.

which would unify a previously scattered structure. One which would operate without competitive interference and whose remit no longer focussed on commercialisation but on the notion of public service broadcasting. The aim of broadcasting for the BBC was to provide programming whose aim at heart was to nurture its citizens on many levels in order to attempt to fulfil what Hendy (2013, 27) describes as the fundamental goal of public service broadcasting, namely, 'advancing human enlightenment.'

COMPETITION

In 1925 a British entrepreneur carried out an experiment which heralded a new era of wireless broadcasting and in doing so introduced a figure who would become a thorn in the side of the BBC for many years to come. Captain Leonard Plugge persuaded Selfridge's department store in London to sponsor a fashion talk which was broadcast from the Eiffel Tower in Paris towards the UK. This transmission, though making practically no impression upon British listeners[17] nor upon the BBC itself, did nevertheless mark the arrival of competition in what was essentially a BBC controlled arena.

The encroachment of competition took the form of broadcasts from foreign stations targeting audiences in Britain and it was not done stealthily. By 1928, when the first experimental broadcasts began to appear with a degree of regularity, the BBC was undoubtedly aware of their activities. The Foreign Director wrote to the Controller in November 1928:

> Hilversum is already transmitting a British advertiser's programmes announced in English, on Sundays, clashing with the Bach Cantatas.[18]

These broadcasts from Hilversum became known as the *English Hour* and worked on the simple principle of British entrepreneurs buying airtime on existing European stations and selling this airtime on to advertisers who would sponsor the actual programmes, thus securing an outlet to commercially exploit their products. Hence, these European stations had segments of their output in English directed solely towards a UK audience.

---

[17] Only three people admitted to having heard the 15-minute broadcast by the actress Yvonne Georges (Woodhead 2012, 204).
[18] BBC WAC: R34/960 Commercial Broadcasting, 1928.

Most of the press was able to report the transmission of such programmes and some sections even in a triumphalist manner:

> The 100,000 listeners in the British Isles—the majority of whom are dissatisfied with the wireless offered them by the BBC—will be interested to learn that... the BBC is faced with the prospect of real and effective competition.[19]

These early, tentative experiments by advertisers failed to make any significant impression on the BBC and as the new decade of the 1930s began, their presence was not perceived as a threat. In an odd way this may not be quite so surprising given that sponsored programmes and concerts from the likes of Hilversum and Radio Toulouse were usually broadcast at a time when the BBC was not actually on-air. Direct competition would possibly have been considered a more genuine threat but such indirect competition, while certainly disturbing for the BBC which considered its position as the sole broadcaster for the UK as absolute, did not at this early stage set alarm bells ringing. The short duration of the broadcasts, their relative infrequency and poor reception may have disguised the true potential of these stations, but figures collated by the BBC at the time on the UK advertising industry were revealing a business model with massive untapped potential. In 1928 UK advertisers bought 9 hours of foreign airtime, in 1929 75 hours and by 1930 this had risen to 300 hours.[20]

In 1929 a number of radio advertising agents got together and formed Radio Publicity Ltd[21] which began transmitting programmes from Radio Paris, already a familiar station in the UK and one easily receivable during daylight hours. This was followed by the International Broadcasting Company (IBC) which was registered as a private company on 18 March 1930. Heading the IBC was Plugge, the man responsible for the Selfridge's broadcast of 1925, the man who would become a not inconsiderable distraction for the BBC for the duration of the 1930s.[22] Radio Normandy began to broadcast IBC programmes from the town of Fecamp in October 1931. The new station's transmitter strength and proximity to the South coast of England meant its potential audience was huge, even larger than its French equivalent, and advertisers found it an attractive publicity

---

[19] *Sunday Dispatch*, 4 November, 1928.
[20] BBC WAC: R34/961 Policy, Commercial Broadcasting, 1945.
[21] Reconstituted as Radio Publicity (Universal) Ltd in 1930.
[22] For a biographical account of Plugge's life and involvement in radio, see Wallis (2008).

machine. While adopting a largely phlegmatic stance regarding Radio Paris and Radio Toulouse, the BBC now began to take notice, and by November 1931 the Director General was writing to the Post Office seeking assistance in 'dealing with the operation of Radio Normandy.'[23]

The BBC's unease with Radio Normandy and the other continental stations broadcasting sponsored programmes in English to the UK was centred around the fact that blatant advertising was taking place and that the quality of the programmes was regarded as inferior when compared to the BBC's own output, an odd complaint but one which typified the BBC's idea of itself at this time as an upholder of quality rather than a chaser of audiences.

Advertising and sponsorship of any form was forbidden for the BBC and it was an area in which it had to tread very carefully. Reith was regularly inundated with complaints from listeners regarding what they saw as possible breaches of the BBC's obligations or anything that looked like it:

> Tonight in announcing Professor Simpson on the wireless his books were enumerated as being published by the Sheldon Press at 2/6d each. Is this advertising or is it not? I thought advertising was prohibited.[24]

Support for the BBC's new found vigour in attempting to construct a case against the continental stations began to come from many quarters including the press who felt equally threatened by the foreign stations' ability to bring advertising to the airwaves, thus possibly diverting a strong source of revenue for the country's newspapers. They subsequently joined forces with the BBC in calling for action by the Post Office and the government, despite the fact that only a few years earlier they had been constantly complaining to Reith regarding on-air mentions of the *Radio Times* and various books and publications (Coase 1950, 101). For the press, which previously saw the BBC, and radio in general, as a competitor in the supply of news, information and entertainment, it seemed working with the BBC could actually be advantageous:

> A practice is growing whereby advertisements in English are broadcast from foreign stations in consideration of money payments. This practice is viewed with great dissatisfaction by the newspaper and also by the BBC.[25]

[23] Letter from Reith to Postmaster General, 16 November 1931. BBC WAC: E2/2/2 File 1.
[24] Letter to Director General from listener John Nayler, Cardiff. BBC WAC: R34/1 Policy Advertising 1924–1939.
[25] Letter from T.W. McAra of the Newspaper Proprietors Association to the Postmaster General, 4 April 1933. BBC WAC: E2/2/2 File 1.

At this point the effects on actual listener figures appeared to be a neg-ligible factor for the BBC and certainly in the early 1930s there is little evidence of an understanding of listening figures or trends. This almost conceited view of the BBC's own position from within the corporation is one which its critics raise again over the course of its future. If anything, the BBC and its supporters were more concerned with the standards of its competitors than any market impact, and a lax approach could be attrib-uted to the attitude of the broadcasting elite towards the endeavours of Plugge and his cohorts:

> I spent part of the afternoon listening to the European broadcast programmes...Radio Paris...I heard two programmes of gramophone records 'sponsored' by an establishment in Brixton Road. A more disgusting display of musical depravity could not be conceived.
> One song was supposed to represent a prisoner praying to his "Gawd in 'eavan" and moaning his "Muvver".[26]

Most of the action taken in the wake of the continental stations becom-ing more established tended to centre around a lot of listening such as described above, followed by a flurry of letters of complaint to the Post Office and the government, but there were also some clandestine opera-tions aimed at gaining an insight into the workings of the companies involved in providing the programmes. There were a number of memos circulating around Broadcasting House at the time with information gleaned from various members of BBC staff who passed on any informa-tion they could uncover regarding the running of the publicity houses:

> Formerly they rented two telephones, but have given one up. The personnel of the office consists of Savi, an Indian or Anglo-Indian, with a young girl to answer telephone calls etc... It is said in Fleet Street that Mr Savi takes on any cause he can get hold of and runs a very low-down type of publicity establishment. It is understood that for purposes of this campaign Savi is financed by Plugge.[27]

Not only were Plugge's broadcasts from Normandy becoming ever more popular in general terms, he was now beginning to transmit special programmes dedicated to particular towns and counties using local speakers

[26] Letter from A.R. Burrows, Secretary General, Union International de Rediffusion to BBC, 10 Oct 1932. BBC WAC: E2/2/2 File 1.
[27] Internal memo from Stephen Tallents, 29 November 1935. BBC WAC: R34/959.

and musicians. Competition was coming in the form of targeted content, and Radio Normandy was not afraid to parade its populist credentials. It was fast becoming a popular radio station for the population in the southern part of the UK. Plugge, also aware that the BBC was complaining to the Post Office, initiated his own campaign called *Hands off Radio Normandy* under a body called the League of Freedom, the aim being to get public opinion on his side by claiming the BBC was using unfair means such as diplomatic channels[28] to curtail the broadcasts that 61% of the British population enjoyed and which were 'sponsored by British firms employing thousands of British workers' (Browne 1985). Plugge was constructing his own populist justification for the broadcasts and was supported by bodies such as the Incorporated Society of British Advertisers which lobbied the government. Radio Normandy was becoming a huge thorn in the side for the BBC from its office just round the corner from Broadcasting House in Hallam Street, but it was no longer the most significant source for concern. Soon Radio Normandy would be overtaken in the popularity stakes by another station based on the continent.

In September 1930, the Grand Duchy of Luxembourg first granted the French company, Societe Luxembourgeoise d'Etudes Radiophoniques, permission to begin broadcasts from the city of Luxembourg. A year later the concession was sold to another French-backed company the Compagnie Luxembourgeoise de Radiodiffusion. The arrival of Radio Luxembourg on the continental radio scene, with its aim of transmitting commercial programmes in English, sent ripples of concern through established broadcasting concerns including the International Broadcasting Union (IBU)[29] whose Secretary General Arthur Burrows[30] wrote to the BBC Chairman in respect of the fact that Radio Luxembourg had constructed a powerful transmitting station, unique in Europe:

> I want to discuss the most diplomatic method of handling the delicate situation which is already foreseen… in connection with the opening

[28] For example, the BBC had approached the Spanish Ambassador to London regarding English language programmes emanating from Spain. BBC WAC: E2/2/2, Advertising in English by Foreign Stations, 1935.

[29] Formed in 1929, the IBU aimed to promote exchanges between European public service broadcasters and mediate in technical disputes. It was the precursor to the European Broadcasting Union (EBU), formed in 1950. In some early BBC documentation, the acronym UIR is used referring to the French: Union Internationale de Radiophonie.

[30] A former BBC newsreader, presenter and Director of Programmes.

of Radio Luxembourg and the efforts which are now being made to obtain control of other European stations for advertising purposes. It looks as though the use of broadcasting stations for advertising purposes will be unavoidable once Luxembourg has begun to shout with a loud voice all over European territory.[31]

Burrows also raised his concerns that the Compagnie Luxembourgeoise de Radiodiffusion was not a member and therefore had no moral responsibility towards the IBU. This, coupled with the fact that Radio Luxembourg's replies had been evasive for some time at various conferences, meant his worries duly proved to be propitious. By the end of 1933 Radio Luxembourg was broadcasting from a new, powerful transmitter which could cover vast swathes of the UK including the north. Not only that but it was transmitting on a long wave wavelength outside the accepted broadcasting bands as agreed at the Prague Conference of 1929.[32] Earlier in the year the BBC had invited a representative from Radio Luxembourg to Broadcasting House on 24 June 1933 to explain their actions. Any sympathy for the BBC was not evident as Monsieur Tabouis explained his company's commercial justification for its operations and outlined its plans for creating an international station which would compete with the BBC.[33] Radio Luxembourg was adamant that it wished to build on its footprint in Britain, competing with the BBC and all other continental stations transmitting to the UK. As Street (2006, 257) notes, its actions made it 'a *pirate* station in a sense that none of the other continental broadcasters were.'

The BBC had tried all sorts of tactics to quell the presence of the continental stations ranging from pressure on the host governments, action through the IBU, direct negotiations with the companies themselves and coordinated attacks alongside the National Press Association but all to no avail. In fact, a newspaper ban on listing the programmes of the continental stations only led to them publishing their own listings magazine known as *Radio Pictorial* which in itself became a major competitor for the BBC's own *Radio Times*. Indirect pressure was also applied by denying the continental stations any facilities or programme material they required. For example, Luxembourg approached the Post Office to request taking the

---

[31] Letter from A.R. Burrows (Secretary General, IBU) to Sir Charles Carpendale (BBC Chairman), 31 December 1931. BBC WAC: E18/283/1.

[32] Meeting of the IBU Technical Committee, Prague, 4 April 1929, to agree on the partition of wavelengths.

[33] BBC WAC: E18/283/2. International Organisations. UIR: Radio Luxembourg.

1937 Coronation Programme for re-broadcast which the Post Office duly denied to them.[34] There was also considerable frustration at hearing BBC Dance Orchestra records being played by Luxembourg.[35]

Another tactic was to apply pressure on the BBC's own broadcasters and artists. Some high profile BBC personalities who broadcast on the BBC and then also offered their services for the studios of continental broadcasters soon found themselves walking a precarious tightrope as was evident in the case of Christopher Stone, often considered to be the BBC's first disc jockey:

> There remains to consider what action we should take especially as Luxembourg is the most notorious of the 'pirates'... There is the view that while this thing is going on Mr Stone's engagements with us should be gradually reduced.[36]

Stone was not the only name to be affected as the minutes of a meeting to discuss Radio Luxembourg's programming reveal.[37] In addition to the decision to not offer Stone any engagements after his present contract expired, it was also decided to hint to Leslie Baily[38] that his activities were counter to the interests of the BBC. This was as a result of the fact that as a journalist, he had been publicising the Luxembourg broadcasts in the *Sunday Referee* which itself was a sponsor of programmes. Also at this same meeting, it was decided that an informal warning be disseminated among variety artists to the effect that they would be prejudicing their position with the BBC by any involvement in broadcasts from Luxembourg. The minutes conclude with the thought that more drastic action ought to be taken against artists such as comedians Clapham and Dwyer, who were primarily BBC artists in that 'their popularity was deemed entirely due to the BBC and no one else.'[39] A formal policy was adopted and staff became aware of it, they could no longer have the best of both worlds:

> Will departmental executives kindly cause it to be known among artists that the practice of the Luxembourg station to transmit programmes in

[34] BBC WAC: E18/283/6. International Organisations. UIR: Radio Luxembourg.
[35] BBC WAC: E18/283/5. International Organisations. UIR: Radio Luxembourg.
[36] Internal memo from Gladstone Murray, 14 August 1934. BBC WAC: E18/283/3 International Organisations. UIR: Radio Luxembourg.
[37] BBC WAC: E18/283/3. International Organisations. UIR: Radio Luxembourg.
[38] Producer who would later be feted for his *Scrapbook* programmes.
[39] Notes from meeting to discuss Radio Luxembourg programming, 29 August 1934. BBC WAC: E18/283/3.

the English language especially for reception by listeners in this country is contrary to British policy and therefore disapproved of by the corporation.[40]

This method of dealing with competitors appeared to be effective as artists and broadcasters now had to decide on which side of the fence they would sit and it had an impact on the business of agents who also had to be careful not to employ moonlighting staff. Indeed the agent responsible for Clapham and Dwyer, the Carlyle Cousins and Tommy Handley reassured the BBC that he would not accept any future contracts for these artists from Radio Luxembourg.[41]

In June 1935 the BBC continued its diplomatic efforts by meeting with representatives from the Government of the Grand Duchy of Luxembourg, and while the latter made clear it had no wish to offend the BBC, it argued that the broadcasts were not in conflict with any IBU resolution as such resolutions were only directed towards the prohibition of political propaganda and not commercial advertising.[42] Radio Luxembourg itself told the BBC that any ban on its activities could not be enforced and, even if it were, the broadcasts would only continue from Normandy or Paris or any one of the other stations and that with the rapid development of short wave transmissions, such advertising could soon be emanating from stations in the USA.[43] Alas, this was a prediction which never materialised but one which revealed the capacity for continued technological innovation to create threats and opportunities.

Having failed in its attempts at employing diplomacy and recourse to international agreements to quell the activities of Radio Luxembourg, and also having warned its own staff of dabbling in broadcasts from abroad, the BBC decided it was time to finally engage in a battle on the airwaves. In the latter half of the decade, Radio Luxembourg was hitting the BBC where it hurt most, namely, at times when the BBC was not on-air and particularly on Sundays, when its audience figures in Britain were greater than for the rest of the week. Reith's renowned strict Presbyterianism meant he believed in upholding the sabbath, so entertainment of any sort could not be heard on the BBC on a Sunday. This became known as the

[40] Internal Memo, 31 August 1934. BBC WAC: E18/283/3.
[41] Memo from Variety Executive, 10 September 1934. BBC WAC: E18/283/3.
[42] BBC WAC: E18/237/7.
[43] BBC WAC: E18/237/7.

BBC's 'Sunday Policy' and was a stance upon which stations like Radio Luxembourg and Radio Normandy nurtured their popularity. Many listeners did not share Reith's view; after all, it was the only day of the week when they actually had time to relax and listen to the wireless in their homes, but what they were offered from the BBC was not the most entertaining of content. On the other hand, the continental stations provided music and entertainment which many saw as perfect ingredients for a Sunday. Reith however, was unwavering in his approach to BBC Sunday programming which constituted a mixture of classical music along with religious talks and readings, but it seemed that a more flexible approach to Sunday programming might be the only route left to take in order to at least counterbalance the hegemony of the continentals in this battleground and discussions began about revising the BBC's output on Sundays.[44]

Initially the BBC can be blamed for continuing its rather aloof approach with regard to its Sunday Policy. Not only was Reith driven by his religious puritanism, but he appeared to be supported by some senior staff keen to pursue an elitist approach to programming:

> On the whole, Luxembourg programmes are presented in such a
> tedious manner and are so monotonous that I think we have less to
> fear from them than we imagine. Concentration on our own standards,
> particularly on Sundays seems to me to be the most promising line of
> approach.[45]

Sunday broadcasts represented a major investment for British advertisers precisely because the target audience was at its largest compared to any other time in the week. In fact 90% of money spent on broadcast advertising by English advertisers around this time was spent on Sunday.[46] The huge commercial success of programmes such as *The Ovaltineys' Concert Party* which went out on a Sunday evening and which 'secured an entire generation of listeners for Radio Luxembourg' (Street 2015, 26), certainly hammered home the need to amend the BBC's output. By 1936 the BBC was committed to a 'lightening' of its Sunday programmes and the Programme Controller, Cecil Graves, formally announced changes in the

[44] Programme Revision, 22 April 1936. BBC WAC: R34/874/3 Policy, Sound Broadcasting, File 2B.
[45] Memo from Entertainment Executive, 10 April 1935. BBC WAC: E18/283/4.
[46] *Shelf Appeal* magazine, October 1935.

construction and scope of Sunday programmes alongside the introduction of more variety in daytime broadcasts and a more attractive arrangement of evening programmes.[47]

The 1936 Ullswater Committee Report[48] into the future of broadcasting had criticised the BBC's Sunday programming for their 'lack of attractiveness' while at the same time recommending steps be taken to stop broadcasts from abroad. A 'lightening' of the Sunday schedule continued over the next number of years although with a sense of duty not to completely vulgarise established Sunday standards and with the only defined parameters being 'no straight dance music and no variety.'[49] Of course, criticism came from many quarters including the established churches, and the BBC had to greet a number of deputations which came to protest.[50]

The BBC also decided to extend its hours of broadcasting during the week by opening earlier in the morning, from 8.00am in fact, something which the continental stations had again instigated. It seems surprising to present-day observers that the coveted breakfast slot was for the BBC borne out of necessity in order to keep pace with its competitors. For the BBC, it was a case of following the lead of the commercial stations as listeners were making their feelings felt:

> My two sons who get their breakfast between 7 and 8 each morning are very glad of the programmes then broadcast from Normandy and we do not in the least mind being told that there are such things as bile beans and Grasshopper ointment. I think we should be disappointed if there was nothing on the radio at that time of the day.[51]

In dealing with the threat to its position vis-à-vis the commercial stations, the BBC eventually developed a more mature approach to understanding its competitors than the clandestine surveillance of offices as hinted above. Instead it came to rely on audience and market research with the setting up of the BBC Audience Research department under

[47] Announcement on Programme Revision, 24 September 1936. BBC WAC: R34/874/3 2B.

[48] Report of the Broadcasting Committee (Ullswater Report), February 1936, Cmd 5091. London, HMSO.

[49] Memo, 27 March 1935. BBC WAC: R34/882/2, Policy, Sunday Programmes, File 2.

[50] Note from Broadcasting House, Edinburgh, 12 April 1938. BBC WAC: R34/882/3, Policy, Sunday Programmes, File 3A.

[51] Letter from listener in Southampton, 13 November 1935. BBC WAC: R34/101.

Robert Silvey in October 1936.[52] Up until this point the BBC's only information on its audience came from listeners' letters. Using statistical analysis, the true impact of competition could now be calculated. Silvey and his staff could produce figures showing that the BBC lagged way behind Radios Luxembourg, Normandy and Lyons on Sundays[53] or that advertising expenditure on the continental stations had more than doubled each year between 1934 and 1937,[54] proving that significant competition was a forceful reality.

By the late 1930s many of the continental stations had become well established, and the threat from a myriad of ever-expanding stations was alive and well. News was constantly reaching the BBC of the activities of Captain Plugge on the continent in his undiminished forays into the boardrooms of foreign stations.[55] The BBC was also probably disappointed to learn of its former chief engineer P.P. Eckersley[56] who could now be heard on-air on Radio Toulouse.[57] As war approached, the BBC's position was looking rather unsteady. Both it and the Post Office had failed to halt the impressive progress of the continental stations. It seemed as though the continued growth of commercial broadcasting was unstoppable, feeding off hungry listeners and insatiable commercial interests. The BBC on the other hand was striving to placate the masses through more appealing programmes while at the same time fending off accusations that it was toying with its public service remit by providing programmes on a Sunday for example, which merely replicated the 'terrible trashy programmes'[58] of the continental stations.

Radio had emerged as a scientific advance before transforming into a broadcasting industry and finally becoming a ubiquitous product. The first wireless receiving set was a crystal set, a very simple radio receiver requiring no direct power source as it functions on the power received

[52] For a detailed account of the workings of the department see Silvey (1974).

[53] Audiences for Foreign Programmes on Sundays, 4 February 1938. BBC WAC: R34/960.

[54] Radio Advertising, 21 October 1937. BBC WAC: R34/959.

[55] Note from Sir Charles Carpendale (BBC Chairman) 16 June 1937. BBC WAC: E2/2/2, Advertising in English by Foreign Stations.

[56] Chief Engineer, British Broadcasting Company, 1922–1927 and Chief Engineer, BBC, 1927–1929 (See Eckersley 1941).

[57] Memo, 4 October 1937. BBC WAC: E2/2/2, Advertising in English by Foreign Stations.

[58] Letter from an unnamed listener to BBC Director of Programmes, 4 June 1938. BBC WAC: R34/882/3, Policy, Sunday Programmes, File 3A.

from radio waves by a long wire antenna. During the mid-1920s the arrival of thermionic valves revolutionised radio receivers. Progress in valve technology meant it became the ascendant mode of listening, despite the fact that valve sets were initially more expensive. By the latter half of the 1920s, however, the price began to fall, and this would lead to the eventual demise of the crystal set by the mid-1930s. Valve sets allowed for communal listening, and so the wireless set found a new and significant place in the home; as Crisell (1997, 17) notes, this marked:

> The transition from radio receivers as unsightly pieces of apparatus, the playthings of
> enthusiasts and eccentrics which had to be hidden in cupboards when not in use, to
> aesthetic objects—pieces of furniture in their own right.

As sets now moved into the living room or parlour, wireless became the boom industry of the 1930s, and every town had shops selling and repairing wireless sets. Radiolympia, or the Radio Show as it was popularly known, was the high point of the radio enthusiast's year held at Olympia in London. Here the radio manufacturers displayed their wares, offering first glimpses of the latest set designs. There were substantial exhibits by the Post Office and of course the BBC, which broadcast variety shows from a theatre built within the exhibition hall. The chance to see radio personalities no doubt contributed to the popularity of Radiolympia, with visitors rising every year since the first National Radio Exhibition, held in 1926, and peaking with attendance figures of 238,000 in 1934. The 1938 Radiolympia was the last pre-war exhibition and at this show the 1939 radio and television sets were displayed. Television had now joined radio in the broadcasting arena.

Experimental television transmissions by the BBC began in 1929, and on 2 November 1936, the Postmaster General formally opened the station at Alexandra Palace where the two competing systems of Baird and Marconi broadcast alternately for two hours a day. After February 1937, the Marconi system only was employed, but even then it was broadcasting to a very small audience within a short radius from its Alexandra Palace base. But with a single television system now in place, consumers could invest in sets without fear that they would become obsolete goods in the near future. At the 1938 Radiolympia, 16 firms displayed television receivers. This was a marked improvement from the 1936 Radiolympia exhibition when only seven radio manufacturers demonstrated television sets. The relative success of the 1938 event marked a shift in BBC policy towards the new medium:

> This year's Radiolympia exhibition is regarded as a clear indication that a hopeful change has taken place in the television situation... Programmes are believed to have proved their entertainment value and reliable sets are now within the reach of a very large public... The BBC considers that a vigorous and sustained effort to market television sets is essential to the success of the service.[59]

This decision to devote resources to promote television as a new medium was therefore based on a number of key factors. Firstly, the technology was considered to have improved to a point where it offered a valuable service. Secondly, it carried content which audiences would embrace. And thirdly, the price of receiving sets had reached a point where successful penetration could be achieved. Such factors would also play a prominent role in the promotion of DAB over 50 years later. However, on the outbreak of the Second World War in September 1939, the television service was closed down. Television's onslaught on radio would come some years later.

Conclusion

A number of characteristics define this early period in terms of the position of the BBC and its competitors and the state of the UK radio industry. An important point is the fact the BBC was born out of competing interests. The British Broadcasting Company emerged out of necessity through cooperation between a multitude of wireless enterprises. This cooperation between competing interests was of course initiated by governmental action, but the most glaring fact is that competitors came together under one umbrella, acting in the best interests of the industry at large. The British Broadcasting Corporation was formed by removing the commercial interests and transferring the organisation into public hands with a public service remit. This was done in an attempt to secure the new business of broadcasting for the British public by providing appropriate programming and upholding certain standards befitting an activity which

[59] Note on Radiolympia Exhibition, October 1938, G109/38. BBC WAC: T16/78, TV Policy.

could yield so much power and influence. Centralising broadcasting in this way meant that listeners were at the mercy of BBC programming and it was the weaknesses in the BBC's programming policy which inspired commercial initiatives by entrepreneurs who recognised how to satisfy listening demands not catered for by the BBC and in a way which could generate profit.

What is striking about the early days of the BBC is its insouciance regarding the continental stations and the lack of any empirical understanding of the audience and the radio industry at large, at least until the latter half of the 1930s. In Britain, the radio industry was created out of a new technology which was moulded by various interested parties into something which it was thought had economic potential. From this came a broadcasting service which subsequently came under attack from home and foreign competitors keen to profit from broadcasting's comfortable relationship with advertising. A number of important characteristics of radio are born in this period: firstly, the creation of a broadcasting industry out of a wireless technology; secondly, a distinction between two forms of broadcasting—a commercial, profit-driven form and a public service form; and thirdly, a determination for competitors to fight for a place within the industry, using whatever means and with a mission to provide populist programming not catered for by the body charged with providing a public service. As the BBC spearheaded broadcasting of what it deemed quality or beneficial programming, then so the commercial competition pioneered lighter, populist programming. The BBC served its audience by installing its national transmitter network, thus bonding the nation under the new medium, and then providing output which it thought best for the nation to consume. The commercial sector on the other hand provided output which would appeal to the largest audience and it was efficient at targeting its audience—instigating breakfast programming is a good example of how the commercial stations could shape the future of radio as much as the BBC.

At the outbreak of the Second World War, one could analyse UK radio as having gone from open experimental playing field, to cooperation necessitated by chaos of the airwaves, to a consolidation of broadcasting in a single body followed by the emergence of a radio market. It is interesting to ponder how the UK radio industry would have evolved had the Second World War not broke out. Certainly, the BBC would have found itself facing a severe crisis had the continental stations continued to operate in the same manner for another decade.

Reith resigned as Director General in June 1938 and by 1939 peace in Europe was crumbling. What would happen next to the BBC and to radio broadcasting in Britain? Radio was a powerful tool of mass information and, as both politicians and entrepreneurs recognised, also a powerful tool of propaganda. As the war began, it was easy to see for which purpose it would be used. Despite the emergence of two tiers within the UK radio industry, the war eradicated the BBC's competitors overnight, and it was to regain its position as the sole broadcaster to the UK audience, although as a result of circumstances beyond which it had any influence.

References

Briggs, A. *The History of British Broadcasting in the United Kingdom. Volume 1: The Birth of Broadcasting.* Oxford: Oxford University Press, 1995.

Briggs, A. and Burke, P. *A Social History of the Media.* London: Polity Press, 2002.

Browne, D. "Radio Normandie and the IBC Challenge to the BBC Monopoly." *Historical Journal of Film, Radio and Television* 5, no. 1 (1985).

Burns, R. *Communications: An International History of the Formative Years.* London: Institution of Engineering and Technology, 2004.

Coase, R. *British Broadcasting: A Study in Monopoly,* London: Frank Cass, 1950.

Coyer, K., Dowmunt, T. and Fountain, A. *The Alternative Media Handbook.* Abingdon: Routledge, 2007.

Crisell, A. *An Introductory History of British Broadcasting.* London: Routledge, 1997.

Eckersley, P.P. *The Power Behind the Microphone.* London: Jonathan Cape, 1941.

Hendy, D. *Public Service Broadcasting.* London: Palgrave Macmillan, 2013.

Hennessy, B. *The Emergence of Broadcasting in Britain.* Lympstone: Southerleigh, 2005.

Hilmes, M. *Network Nations: A Transnational History of British and American Broadcasting.* Abingdon: Routledge, 2012.

Lyons, E. *David Sarnoff: A Biography.* New York: Harper and Row, 1966.

Marconi, G. "Improvements in Transmitting Electrical Impulses and Signals, and in Apparatus therefor." London: GB Patent Office, Number: GB12039, 2 June 1896.

Reith, J. *Broadcast Over Britain.* London: Hodder and Stoughton, 1924.

Reith, J. *Into The Wind.* London: Hodder and Stoughton, 1949.

Sarkar, T., Mailloux, A., Oliner, A., Salazar-Palma, M. and Sengupta, D. *History of Wireless.* Hoboken: John Wiley & Sons, 2006.

Sarnoff, D. *Looking Ahead: The Papers of David Sarnoff,* New York: McGraw Hill, 1968.

Scannell, P. and Cardiff, D. "Serving the Nation: Public Service Broadcasting Before the War." In *Popular Culture: Past and Present*, edited by Waites, B., Bennett, T. and Martin, G. London: Croom Helm, 1982.

Scannell, P. and Cardiff, D. *A Social History of British Broadcasting 1922–1939*. Oxford: Blackwell, 1991.

Scott, P. "The Determinants of Competitive Success in the Interwar British Radio Industry." *Economic History Review* 65, no. 1 (2012).

Silvey, R. *Who's Listening? The Story of BBC Audience Research*. London: George Allen and Unwin Ltd, 1974.

Smith, A. *British Broadcasting*. Newton Abbot: David and Charles, 1974.

Street, S. *A Concise History of British Radio 1922–2002*. Tiverton: Kelly Publications, 2002.

Street, S. *Crossing the Ether: British Public Service Radio and Commercial Competition 1922–1945*. Eastleigh: John Libbey Publishing, 2006.

Street, S. *The Memory of Sound: Preserving the Sonic Past*. Abingdon: Routledge, 2015.

Wallis, K. *And the World Listened: The Biography of Captain Leonard F. Plugge*. Tiverton: Kelly Publications, 2008.

Woodhead, L. *Shopping, Seduction and Mr Selfridge*. London: Profile Books, 2012.

Young, F. *The Trial of Hawley Harvey Crippen*. London: William Hodge and Co, 1920.

CHAPTER 3

Wartime and Post-War Radio Broadcasting: BBC Hegemony and Commercial Sector Hiatus

The period of the Second World War gave the BBC time to consolidate its output for the duration of the war and also prepare its radio strategy for the post-war era when opposition was likely to return, as all commercial opposition disappeared from the airwaves at the outbreak of hostilities before slowly regaining a footing once the war had ended. The BBC could now continue to champion its own considered standards in broadcasting but would also need to make headway in delivering content which might appeal to the significant audiences which had strayed towards competitors throughout the 1930s. Achieving a comfortable balance between these two directions might represent a great challenge for the BBC, but the question of how far to go along the path to populist radio would also be challenging if the BBC wished to maintain its standards while at the same time taking the necessary steps to provide a service which could compete. As well as radio, the post-war era witnessed the revival of television which would, in the forthcoming years, become more of a threat to BBC Radio than any commercial radio activity and indeed would similarly have a detrimental effect on the commercial radio sector itself. By the end of this period, television would have superseded radio as the primary medium, but radio—both BBC and commercial—would still be holding its ground, albeit on shaky foundations.

© The Author(s) 2018
JP Devlin, *From Analogue to Digital Radio*,
https://doi.org/10.1007/978-3-319-93070-1_3

Outbreak of War: The BBC Home Service

At 11.15am on Sunday 3 September 1939, the country tuned into the BBC to learn from Prime Minister Neville Chamberlain that Britain was now at war with Germany.[1] Chamberlain's words from the Cabinet Room at 10 Downing Street last no more than five minutes and are followed by government announcements read by Alvar Lidell, warning of impending air raids. Although most of those listening on that Sunday morning were not surprised by the actual declaration of war, the fact that the whole country was able to listen to such a momentous declaration, live and at the same time, lent a greater degree of poignancy to the words.[2] The gravity of the broadcast was also felt by the speaker as it took a further 48 hours before Chamberlain himself could write 'war declared' in his own diary (Dutton 2001, 57).

The content of the broadcast was not a surprise simply because government bodies, businesses, various institutions and the population at large had been expecting it and had even been making plans for the looming hostilities ranging from how animals would be dealt with to the distribution of gas marks. Plans for broadcasting had been afoot as well in the preceding months both inside and outside the BBC so that on 1 September 1939, two days before Chamberlain's broadcast, some BBC departments were moved out of London to various locations outside the capital, including the move of the Light Entertainment department to Bangor in North Wales.[3] Also, all domestic broadcasting became limited to a single national wavelength which was introduced to assuage 'fears that transmitters might be used as beacons for enemy aircraft' (Haley 1946). During the course of the 1920s and 1930s, the BBC had been offering two domestic services, the National Programme and the Regional Programme; both were now merged into a new network, the Home Service.

People turned to the BBC and to the nascent Home Service as there was really very little alternative after this point. But for those seeking the type of entertainment provided by the BBC or its competitors in the previous years then they were to be disappointed. Programme content on the new Home Service consisted largely of a staple diet of news, information and gramophone records. While the public was of course hungry for

[1] http://www.bbc.co.uk/archive/ww2outbreak/.
[2] Although Lewis (2013) argues this speech has now become a cliché in media histories of the time.
[3] Other locations included Evesham, Bristol, Oxford, Manchester, Glasgow and Bedford.

information, the lack of entertaining output left many thinking 'broadcasting as they had known it had come to an end' (Bridson 1971, 72). Not only that but with the closure of theatres, cinemas and concert halls, the average citizen was suffering from a dearth of genuine entertainment. Criticism came from listeners and the press and it continued to rise over the autumn of 1939. It would appear that the BBC, which had evidently gained an understanding of audience needs during the years of intense competition from Radio Normandy and Radio Luxembourg, was intent on eschewing any change to its programme output despite now having an open opportunity to consolidate its position, as Nicholas (1999, 63) observes:

> The conventional view of the BBC is of cumulative 'democratization'
> as the BBC became more obviously responsive to listeners' tastes.
> But even proponents of this view have underlined how contrived,
> patronizing and out of touch many of the BBC's 'people's' programmes
> were.

By 16 September, when the expected air raids had failed to materialise, the government agreed to relax the compulsory closure of theatres, cinemas, dance halls and places of public entertainment by allowing certain venues to re-open including some West End theatres, some of which triumphantly remained open for the entire duration of the war.[4] The same happened with cinemas. After initial universal closure, many were re-opened again as a way of boosting morale and providing a principal source of recreation for a nation at war. If cinema and theatre were seen as essential requisites for maintaining morale, then surely a similar role could be envisaged for radio, particularly as it was even more accessible to a greater number of people. By 1939, 73% of UK households owned a radio licence suggesting a total potential audience of perhaps 35 million out of a population of 48 million (Nicholas ibid, 64).

At a governmental level, Chamberlain and his cabinet considered shutting down BBC transmissions completely in the event of hostilities, believing they 'had no part to play in modern war and should cease as soon as war broke out' (Briggs 1995b, 71). However, the value of broadcasting as a tool of mass communication in providing information and boosting

---

[4] The Windmill Theatre remained open for the duration and coined the enduring motto: 'We Never Closed.'

morale soon became obvious, and this notion was not pursued. It must be stressed however that British radio did not become a tool for propaganda in the way it did in Germany or the USA. The BBC was reluctant to consider any deliberate perversion of the truth in order to maintain national morale, considering such a step 'not in accordance with BBC policy.'[5] Briggs (ibid, 10) notes that this BBC emphasis on truth produced 'long-term dividends in developing a reputation for impartiality which persisted in the decades following the end of the war.'

Finding itself in the unique position of being in a virtual radio monopoly position meant the BBC could now take steps to ensure that this position could be sustained into the long term. As Briggs (1985, 186) notes, a fear of commercial interests as well as a desire to entertain troops serving abroad 'galvanized the BBC' and support for this stance came from leaders of both the British and French military. On 29 November 1939, the BBC Control Board agreed on plans for a new special service for the 'fighting forces,' and at this stage it was referred to as the Forces Programme. Far from the core home audience driving change, it was instead the troops abroad, and their needs, which was the audience sector that instigated a significant amendment to programming.

OUTBREAK OF WAR: THE BBC FORCES PROGRAMME AND GENERAL FORCES PROGRAMME

Just as the home audience had been getting used to the *phoney war*, so also had the soldiers across the channel who found themselves living peacefully among the French rather than fighting the Germans, and radio was seen by military leaders as a way of relieving the boredom of this period. After the Control Board had agreed that plans should be made for a 'special service for the forces,'[6] the Board of Governors subsequently decided that following the example shown by Radio Normandy, the erstwhile BBC Sunday Policy should not be pursued in the planning of the new service.[7] Following internal discussions, a report called Broadcasting to the Troops in France was prepared in December 1939 which concluded with the need for an almost exclusively light service for the troops in France, but importantly, one which would also serve a domestic audience.

[5] Home Service Board, Minutes, 3 November 1939. BBC WAC: R3/16/1.
[6] Control Board, Minutes, 29 November 1939. BBC WAC: E4/11.
[7] BBC Board of Governors, Minutes, 22 December 1939. BBC WAC: R1/1/3.

The BBC Forces Programme was launched on 7 January 1940, a service very much modelled on Radio Normandy and the result of intensive debate and research into the performance of the latter in appealing to both the audience in Britain and among forces in France over the previous four months. Although primarily aimed at the British Expeditionary Force, the wavelengths for the Forces Programme were shown in the *Radio Times* and its programmes could be heard relatively easily on wireless sets in Britain. It featured drama, comedy, popular music, features, quiz shows and variety as well as the obligatory news bulletins, information programmes and talks, and as Street (2006, 189) notes, by the middle of the war the Sunday page in the *Radio Times* was showing home audiences 'a very different bill of fare to that which they had been used to in peace time.' The strict Sunday Policy was cast aside with the first broadcast of the variety show *Garrison Theatre* on Sunday 18 February 1940 which Baily (1968, 157) argues was 'the beginning of the end of treating Sunday as different to other days.' Within weeks, output on the Forces Programme, particularly on Sundays, began to reflect popular audience tastes. So we see the appearance of what are now deemed radio classics such as *Variety Bandbox* (1941), *Workers' Playtime* (1941) and *Music While You Work* (1940). Indeed Nicholas (1999, 71) argues that:

> The implications of the establishment of the Forces Programme were far-reaching, though few within the BBC were prepared to admit as much... it marked the most decisive break yet with the BBC's Reithian past.

By the early 1940s there was no commercial competition beyond a competitive element emanating from Germany featuring the propaganda broadcasts of Lord Haw-Haw[8] whose broadcasts from Hamburg with the chilling introduction 'Germany Calling, Germany Calling' could be picked up across most of the UK. It is thought that on average six million people listened to each of his broadcasts (Cole 1964, 127), and while some found the broadcasts so absurd that they were seen merely as a way of relieving the tedium of life in Britain during the war, the broadcasts also provided the British public with information which had been censored at home.[9]

---

[8] Lord Haw-Haw was the nickname given to William Joyce who broadcast German propaganda to Britain between 1939–1945 (See Kenny 2003).

[9] Hamburg Broadcast Propaganda: The Extent and Effect of its Impact on the British Public During Mid-Winter 1939/40. 8 March 1940. BBC WAC: R9/9/4.

There is no doubt that Lord Haw-Haw, as well as putting fear into the minds of some listeners, also provided a form of entertainment and information and a diversion from the BBC news bulletins. Indeed, a BBC report from January 1940 showed that in the previous month, 30% of adult listeners were tuning into Hamburg.[10] Some listeners thought that the best way to deal with Lord Haw-Haw was to treat him as entertainment.[11] In fact, there were protests from the audience when George Formby was put on air at the same time as a Lord Haw-Haw scheduled broadcast in April 1940.[12]

Achieving the appropriate tone of the Forces Programme was still proving unattainable for the BBC during its first few weeks, as attested by an internal report which revealed that:

> The troops preferred light music and disliked plays and talks. They tolerated news, but thought announcers, while respected for their authority, could do with being more lively.[13]

Nevertheless Nicholas (1999, 71) believes the Forces Programme eventually played an important role in the BBC's understanding of radio in that it:

> ... represented the BBC's acknowledgement that one of the functions of radio was to provide light background entertainment for the casual listener.

Nicholas also notes how a programme service aimed at British forces abroad found a presence in the listening habits of the domestic audience and points out that by 1941, 60% of the civilian public was regularly tuning into the Forces Programme in preference to the Home Service. Among what Nicholas refers to as this 'ordinary listening public,' she highlights working class housewives:

> ...whose existence as a significant component of the daily audience was one of the BBC's more striking wartime discoveries. (Nicholas ibid).

[10] BBC Interim Report: The Effect of Hamburg Propaganda in Great Britain. January 1940. BBC WAC: C4/40.

[11] BBC Memo from Graves to Nicolls, 12 March 1940. BBC WAC: R28/121/1.

[12] BBC Home Service Board Minutes, 5 April 1940. BBC WAC: R3/16/1.

[13] BBC Report; Ryan, A. Listening by the British Expeditionary Force. 23 January 1940. BBC WAC: R1/8/1.

Such discoveries began to alter BBC output, and when the Forces Programme was renamed the General Forces Programme from 27 February 1944, it continued to maintain its distinction from the Home Service with a more populist schedule and more relaxed style and was now officially available to both the domestic audience and the allied forces arriving in Europe. It is fair to say that the influx of Americans to the European theatre of conflict around the latter part of the war had a huge effect on British radio by lightening it even further. The American GIs complained about the BBC, they found 'the music dull, the humour alien and the announcers stiff' so much so that their response was to set up their own radio network (Morley 2001, 2). Hence, the American Forces Network (AFN)[14] sought to provide a form of entertainment which American servicemen had been used to at home and this had the effect of impressing the indigenous population too:

> American musical influence was most profoundly felt through the presence of U.S. troops in the country... and the programmes of the American Forces Network. (Barnard 1989, 28)

Despite the popularity of the AFN, BBC Director General Robert Foot and Editor-in-Chief William Haley[15] believed that the new element of choice in the BBC's offerings should 'cause what present demand there is for commercially provided competition to subside.'[16] This reveals that the BBC was still wary of the possibility of a commercial competitor arriving on the radio landscape at any time and particularly at this late point in the war. It is reasonable to suggest that the broadcasts of the AFN were a sharp reminder of the commercial players who were no longer around. They also served to alert the BBC to the fact that its own more populist output did not perhaps go quite far enough to diminish any possible future competitor threat. Street (2002, 82) argues that the AFN was something which had not been experienced before in Britain and 'gave the listener at home a tantalising sense of the less formal style of American broadcasting' something with

---

[14] Radio service for American forces originally formed by the US War Department as the Armed Forces Radio Service (AFRS) on 26 May 1942.

[15] Robert Foot, Joint Director General of the BBC (with Cecil Graves), 1942–1943, Director General 1943–1944. Sir William Haley, Editor-in-Chief, 1943–1944, Director General of the BBC, 1944–1952.

[16] Note by Director General & Editor-in-Chief, 3 January 1944. Post-War Planning, BBC WAC: R34/578/1.

which the BBC struggled in its attempt to eschew an *Americanisation* of British media and it would inform events in British radio 20 years later when the pirate radio stations adopted American broadcasting techniques. But in the meantime the BBC had to address the question:

> How might British radio develop its own popular forms of expression, based firmly in British popular culture, that might resemble American practices somewhat but would not therefore be intrinsically, and dismissably, 'American'? (Hilmes 2003)

When Operation Overlord, the Allied invasion of occupied Europe, began on 6 June 1944, it was felt by the Allied governments that a joint service of entertainment, news and information for the fighting troops would be a better use of resources than providing separate services for American forces (AFN), British forces (BBC) and Canadian forces (Canadian Broadcasting Corporation, CBC). A combined station, called the Allied Expeditionary Forces Programme, operated by the BBC at the request of General Dwight D. Eisenhower,[17] began broadcasting on 7 June 1944, the day after the Normandy landings, on 514 metres (583kHz), providing a service 'dominated by cabaret and swing music' (Baade 2012, 189). The station closed soon after VE Day on 28 July 1945 when the British Forces Network, AFN and CBC re-established their own services in the areas each force was occupying. The following day, the BBC Light Programme began.

The BBC's decision to cater for British forces had the effect of aligning its programming more in relation to what the commercial stations had been offering before the outbreak of war. By 1945 the corporation had changed the style of content for which it was renowned through the introduction of a more vigorous tradition of speech and humour, one that was 'closer to the music hall tradition than the well mannered "variety" of pre-war programmes' (Curran and Seaton 2003, 144). Whatever the contribution of the Forces Programme to troop morale and to morale at home, it was also a good force for the BBC, as along with the Home Service and the Empire Service, it provided the necessary soundtrack to the difficult war years. It also produced a lighter alternative that would survive the war, and this was done largely by embracing a broadcasting model which had been developed by the commercial sector.

---

[17] Supreme Commander, Allied Expeditionary Force.

OUTBREAK OF WAR: THE COMMERCIAL SECTOR

Attempts by the BBC and the Post Office to curtail broadcasts from continental Europe, directed specifically at the British audience, were consolidated by publication of the Ullswater Report in 1935[18] which, in considering 'the constitution, control and finance of the broadcasting service in this country,' proposed that foreign broadcasting should be 'discouraged by every available means.' By the end of the 1930s, a plethora of stations were available to British listeners such as Radio Lyons, Radio Paris, Radio Toulouse, Radio Athlone, Radio Hamburg, Radio Rome and Radio Berlin, but none had such a hold on the British public as the two main players—Radio Normandy and Radio Luxembourg. Such was the impact of Radio Luxembourg in Britain that in the months before the outbreak of war the Foreign Office thought it would be better to try and make use of Radio Luxembourg than seek to destroy it. So, in September 1938, the unprecedented step was taken by the British government of providing Radio Luxembourg with recordings of Chamberlain's speech on the Munich Crisis, and the BBC was to help with the re-broadcasting of further speeches (Pronay and Taylor 1984). Briggs (1995a, 341) suggests the feeling that war might be imminent was changing old relationships and that there was a feeling on the part of the government that Radio Luxembourg should be consulted with, given that Europe was now becoming more and more dominated by propaganda from all angles and that radio was an important tool in its dissemination. Although this could be seen to signify a mellowing of stance towards the commercial sector and even a step towards bridge building, it must be remembered that this position was instigated by the Foreign Office and not by the BBC itself and if this is seen as cooperation then it must be seen as at the express wish of 10 Downing Street. As Deputy Director General Cecil Graves noted, there was a 'desire on the part of the government that Radio Luxembourg should be used.'[19]

The declaration of war however changed everything. Where commercial radio existed it was expected to 'defer to the official voice' (Hobsbawm 1994, 196). Quite simply, the war put the commercial stations out of action. At a single stroke, the Luxembourg government closed down

---

[18] Report of the Broadcasting Committee, 1935 (Ullswater Report), London: HMSO.

[19] Vinogradoff, I., 1945 *History of English Advertising Programmes Broadcast to the UK from Foreign Stations Down to the Outbreak of War.* BBC Secretariat Document. BBC WAC: R34/961

Radio Luxembourg on 21 September 1939 in order to protect that country's neutrality, although it continued domestic broadcasts within Luxembourg until Germany invaded the country on 10 May 1940 and took over the station, using it to relay Lord Haw-Haw's broadcasts from Hamburg. Radio Normandy had one of its transmitters at Louvetot requisitioned by the French authorities, but with the Fecamp transmitter still in its hands, the IBC quickly initiated plans to reinvent itself as Radio International, continuing to broadcast from Fecamp on 212 metres. Leonard Plugge led the IBC campaign to court British government support for running Radio International as a broadcasting service for British forces who began arriving in France throughout September 1939, although to no avail. The IBC then turned its attentions to the French government to keep the Fecamp transmitter in service for the allied troops, a step which was met more favourably (Plomley 1980, 165). On the ground, the IBC began distributing a programme sheet called *Happy Listening* to the British Expeditionary Force troops in the hope that the association with Radio Normandy would stimulate interest and of course the fact that Radio International could also be picked up in Britain, thus continuing the association with Radio Normandy, meant listeners could still listen to familiar programmes and presenters.

In addition to the BBC's unease with the function of Radio International, objections to it were emanating from within France due to fears the Fecamp transmitter would act as a beacon for German aircraft and there were concerns the IBC was operating broadcasts in German as well. Radio International eventually closed on orders from the French government on 3 January 1940. During the relatively short four months of Radio International's existence, it had quickly re-implemented its broadcasting recipe of populist programming which had already brought it so much success and which it used to make headway on a more conservative BBC. Despite an enhanced corporate-wide understanding of competition, the BBC of the 1930s had failed to challenge the commercial broadcasters by amending its output to cater for the tastes of those attracted to the latter's output. Indeed, Street (2006, 190) claims this resistance to change in programming was unshakeable for so long largely due to the 'spiritual presence' of John Reith even after his retirement in June 1938.

The period from September 1939 to January 1940 became almost the 1930s in a post-Reithian microcosm. BBC research from November 1939 showed clearly that the corporation was losing out to other broadcasters with more than half of those whose opinion was sought agreeing to listening to foreign stations.[20] This research was quite startling in a period when one might have presumed the public would have been satisfied with the highly informative and formal output of the Home Service. Instead, it merely highlighted the fact that programming of lighter content was what the audience also desired. But of course it was not just the domestic audience that needed this attention. Radio International's prominence among British troops in France had created an enviable bond between home and abroad centred around a radio station. When Radio International stopped broadcasting on 3 January 1940, the BBC began its experimental Forces Broadcasting only a few days later on 7 January.

The BBC: New Post-War Portfolio

The success of the Forces Programme during the war meant that something of its character and appeal would have to be retained in the post-war era if the BBC was to maintain its position with the domestic audience in peace time. Something would have to sit alongside the Home Service, which continued to preserve the traditional BBC model of broadcasting and had changed little since its pre-war incarnation as the National Programme, and which would continue in that same mould into the future. But the necessity to provide a service in the shape of the Forces Programme was paramount because of the fact that the audience desired such a service, and also because the BBC now had, as a result of its wartime monopoly, secured its position as the sole provider of radio in the UK, a position which in peace time might come under threat should the competitors from continental Europe make a return, or should opponents of the BBC's monopoly gain greater voice in their calls for more competition in broadcasting at home.

Discussions within the BBC regarding the post-war radio landscape had started early. Under Director Generals Foot and Graves,[21] post-war planning had begun as early as January 1941 when the Control Board first began to mention post-war organisation and the potential of one wavelength carrying a 'Forces Programme/Radio Luxembourg type of

---

[20] Silvey, R.J.E., 17 November 1939, The Public's Attitude Towards the BBC and the Extent of Listening to Foreign Stations. BBC Internal Memo. BBC: WAC R34/960.

[21] Robert Foot and Sir Cecil Graves were joint Directors General 1942–1943. Robert Foot was Director General 1943–1944.

material.'[22] By the end of 1942 thoughts about the future became formalised with the first of the Post-War Planning Meetings which considered how the BBC should take the initiative on the cessation of hostilities, how it should face potential competition from the possible re-emergence of sponsored programmes and also what impact developments in radio engineering such as frequency modulation (FM)—which had resulted from impetus given to research by war—might have.[23]

By 1943 Foot was adamant that the BBC had to be prepared for the post-war period, and from then on the BBC began to concern itself with its future role with greater vigour. Briggs (1995c, 30) in fact argues that between 1943 and 1945 'far more time and energy were devoted to planning for broadcasting after the war than had been devoted in 1939 to planning for broadcasting in war-time.' Three facts were evident on the post-war horizon. Firstly, the BBC had a head start should any competitors enter the broadcasting arena, since it held a virtual monopoly position in the UK. Secondly, the Home Service, although not providing the range of programming for all interests, continued to represent a bastion of core BBC values and ought to target the entire UK population with a 'degree of culture which should be that which can be readily assimilated by ordinary people who care even a little about such things.'[24] And thirdly, the success of the Forces Programme was envisaged as a successful broadcasting model which should be consolidated in the form of a 'civilian programme of light character.'[25] It was a Senior Controller, Basil Nicolls, who sketched out in 1944 a scheme for a general 'Home Service', a 'Light' Programme and an 'Arts Programme,'[26] and in doing so he also introduced the novel idea of these new BBC services not just being able to compete with external forces but also with each other; 'the objective is to allow the freest possible competition within the BBC's monopoly.'[27]

[22] Note from S. de Lotbiniere (Outside Broadcasts) to Sir Cecil Graves (Deputy Director General), 1 January 1941. BBC WAC: R34/576, Post-War Planning.

[23] Post-War Planning Meeting minutes, 13 November 1942. BBC WAC R34/578/1.

[24] Editor-in-Chief, William Haley: The New Home Service, 18 February 1944. BBC WAC: R34/417, Home Service.

[25] Director General, William Haley, speech on Post-War Broadcasting at the Radio Industries Club Luncheon, 28 November 1944. BBC WAC: R34/580, Post-War Planning.

[26] Basil Nicolls: Post-War Home Programme Set-Up, 21 December 1944. BBC WAC: R34/420, Home Services.

[27] ibid.

Foot and Haley had already issued a statement in February about the future of the BBC which stated that it should be 'responsible for all broadcasting in the UK', that it should be 'protected against commercial radio' and that the only element of competition 'should come within the corporation itself.'[28] Nicolls took a step further by outlining the nature of the services and emphasising how they would operate as competing forces within the BBC. Briggs (1995c, 51) argues the influence of Nicolls on the making of the new broadcasting structure was considerable and indeed Nicolls' model continues in its basic format for BBC Radio to this day.

The policy of three services gained momentum throughout 1944 and were discussed at a BBC Board Meeting in August 1944 when Haley stated that the needs of the home audience 'had been sacrificed during the course of the war and that greater choice should be made available after the war.'[29] He outlined the three programmes which the BBC wished to provide. The first would be a 'continuation of the existing Home Service, a home programme capable of separating up regionally.' The second would be a 'lighter programme, based on the Forces Programme, carried on Long Wave and thus also available to troops abroad.' And, the third would be of a 'high cultural level, devoted to the arts, serious discussion and experiment.'[30] The Board agreed that all three should be made available as soon as possible after the end of hostilities, although the third might be delayed if the necessary wavelengths were not available. As the war in Europe came to an end on 8 May 1945, the BBC set in motion its plans for the three new networks. The *Radio Times* announced on 27 July 1945 that 'the BBC had served the nation at war and that it would do so as energetically and as imaginatively through the years of peace.' Two days later on 29 July, peacetime broadcasting resumed with the Home Service continuing as it was, only with six new regional services. The General Forces Programme was renamed the Light Programme and an announcement was made regarding a future Third Programme.[31] The country was beginning to return to some degree of normality after a long six years and

[28] Foot and Haley, statement on the Future of the BBC, February 1944. BBC WAC: R34/420, Home Services.
[29] BBC Board Meeting minutes, 10 August 1944. WAC R34/420, Home Services.
[30] ibid.
[31] Programme Changeover Announcement, 29 July 1945. BBC WAC: R34/420, Home Services.

the BBC thought it was too by offering a new range of choice, catering for all tastes and satisfying those desperate for more than one single station in what would become a 'grey post-war cultural landscape' (Martin 2000, 44). Normality was returning to Broadcasting House in a literal sense too with the removal of the heavy concrete fortifications around the building only weeks later (Reid 1987, 80).

While the Home Service was to continue in the same mould as before, it was the Light Programme which was to be the BBC's standard bearer for the type of programming which had been familiar to listeners of the Forces Programme or the pre-war commercial stations. Its remit was unequivocal in its embracing of the characteristics of these latter stations. It was to:

> … interest listeners in life and in the world around them without at any moment failing to entertain them… meaning a strong foundation of entertainment programmes which are used to support our more serious offerings… the title 'Light' does not necessarily mean that the programme is lowbrow, but denotes that it is aimed at those who require relaxation in their listening.[32]

Accordingly, a number of popular programmes became staples of the Light Programme, such as *Family Favourites (1945)* (successor to *Forces Favourites*), *Housewives' Choice* (1946), *Woman's Hour* (1946), *Dick Barton* (1946) *and Mrs Dale's Diary* (1948). However, the new service should not be seen as completely 'light' in content and was not devoid of more serious material as is evident from *Radio Newsreel* (1947) which transferred from the BBC Overseas Service. Certainly the format of *Radio Newsreel*, which 'included lighter and more entertaining news items' (Chignell 2011, 61), suited the service and provided an appropriate level of news for its audience, something the BBC wished to continue despite the fact that the audience appetite for news fell significantly after the war and never returned to wartime levels (Curran and Seaton 2003, 144).

The Light Programme was successful. By October 1945, 51 out of every 100 home listeners were listening to it.[33] One year later the listening figures had increased by a fifth. By the late 1940s regular listeners were listening for an average of nine and a half hours a week as against seven hours a week

---

[32] Memo from T. Chalmers, Acting Controller, Light Programme to J. Langham. 13 May 1948. BBC WAC: R34/454/2, File 2A, Light Programme.

[33] The remaining 49 were listening to the Home Service.

in the case of Home Service listeners and three hours in the case of the Third Programme.[34] The Light Programme audience was acquiring some of the characteristics of the mass television audience of a far later date; in fact Briggs (1995c, 65) claims one can trace a direct line between the Forces Programme, the Light Programme and the first BBC mass television audiences. However, it should be noted that as the fully national Light Programme extended its range and popularity, the Home Service, with its regional variants, continued in its role as the staple of the BBC. Lewis and Booth (1989, 78) reflect that there were still 'strong elements of pre-war paternalism' in evidence with the Home Service, as the BBC continued to dictate what it felt to be appropriate listening for the audience.

The Third Programme went on air on 29 September 1946. Its aim was to fulfil Haley's vision of a programme having a 'high cultural level, focussing on the arts, serious discussion and experiment.'[35] It had been late in starting due to technical issues, namely, finding an appropriate wavelength to allow as wide a coverage as possible. Russia had laid claim to the 514 medium wave and used it to set up a new station, *Soviet Latvia*. Initially the BBC decided to keep the new Third Programme on 514 metres but at a greatly reduced power in order to avoid interference from Riga (Pawley 1972, 327). It was thought the best way of extending and improving its transmissions would be to adopt FM radio transmission which had been developed in the USA in the 1930s for military purposes and had been used during the war,[36] and it had been demonstrated to the BBC Board of Governors as a potential new radio technology in October 1944.[37] The Board concluded it would take a considerable amount of time to persuade the public to buy the necessary receivers or adaptors in order to receive FM transmissions, and building new transmitters would also be expensive, so it reluctantly decided to discard FM at this time and continue to broadcast the Third Programme on the existing 514 medium wave. BBC engineers

[34] BBC: A Review of Listener Research Findings. December 1949. BBC WAC: R34/454/2, Light Programme.

[35] Haley article on the new BBC services. *Radio Times*, 27 July 1945.

[36] FM changes or modulates the frequency of the unmodulated signal while keeping the amplitude of the signal constant. When the frequency is modulated, music or talk is transmitted via the carrier frequency with the effect of improving the fidelity of the radio signal. The distance range for FM transmissions are much more limited than AM, usually less than 100 miles, but as they operate in a much higher range of frequencies than AM radio, this provides for a much clearer sound.

[37] BBC: Board of Governors Minutes, 4 October 1944, BBC WAC R1/12/1.

then began to construct nationwide makeshift masts to boost coverage on the spare 203 metres. This technical dimension to the launch of the Third Programme illustrates a reticence on the part of the BBC to trial a new network on a new platform and a certain caution in approaching a new technology which would require mammoth consumer participation in purchasing new receiving sets. It seems that at this point in the BBC's history such a venture was considered too risk laden. Following its uneasy birth, the Third Programme established itself as a provider of highbrow culture: music, drama and discussion. It came under fire however from some sections of the press for being 'obscure, dull and pretentious' with some asking 'how many people actually listened to it?' (Carpenter 1996, 72). Others argued that what mattered was 'not the size of the audience but the quality of the programmes' (Carpenter ibid, 80) for those interested in its output. The Third Programme's future would continue to be characterised by such debate, but it did achieve Haley's idea of providing a service for a minority audience, an ambition shared 50 years later when the BBC decided on which audiences to accommodate on its new digital radio services.

Haley had envisaged a listening ratio for the Home, Light and Third of 40, 50 and 10[38] although a fundamental principle behind the tripartite system at the start was that each service was to be in competition with the others. This internal element of competitiveness was however abandoned in 1948 as the corporation set out to invoke the 'overall interests of the BBC' when competition between internal departments became too keen (Briggs 1995c, 79). In fact, the idea of competition between networks and their supply departments was to disappear completely at the BBC from this point until it was re-introduced in the form of *Producer Choice* in the 1990s. What the tripartite system does illustrate however is a very early recognition of the notion of 'niche broadcasting' on a very basic level, and this may be seen as marking the start of 'a system of broadcasting which would evolve along cultural strands, culminating just over twenty years later with the creation of Radios 1, 2, 3, and 4' (Street 2006, 192).

RETURN OF THE CONTINENTAL STATIONS

Much of the BBC's proactive policy-making in preparation for the end of the war was centred around consolidating its extant broadcasting monopoly and taking steps to secure that position by addressing audience needs,

[38] Haley, Address to the General Advisory Council, 29 October 1947. BBC WAC: R34/420, Home Services.

catering for those needs and thus denying a possible opening for any budding entrepreneur or any established organisation to gain a footing in the domestic British radio arena. After all, as Haley noted, 'today is still the hindsight of a closing era and tomorrow is the threshold of a new age.'[39] The BBC had set in motion its new tripartite structure, and support from within the Labour government for maintaining the BBC's position was also forthcoming. When the BBC charter came up for renewal in 1946, Prime Minister Clement Attlee argued that domestic and international issues would be at stake and he hoped that the BBC would remain 'untouched from competition' (Briggs 1995c, 33). But what was the status of Radio Normandy and Radio Luxembourg immediately after the war and were they intending to re-enter the radio market?

For Radio Normandy it was unlikely it would return to the airwaves since the French government desired that radio would now be used as a tool to 'serve the interests of the post-war state' and broadcasting was confirmed as 'a state monopoly with public service goals' (Kuhn 1995, 90). This meant Radio Normandy would not be able to have any base on French soil, and the same would apply to Radio Lyons, Radio Toulouse and Radio Paris. Any business interests considering investing in a foothold in France in order to broadcast to Britain found their entrepreneurial endeavours curtailed at French governmental level. Despite this, the BBC never stopped expecting Leonard Plugge or the IBC to suddenly reappear in some form or another. Even in July 1946 suspicions could still be aroused as seen in an internal memo from the time:

> I got some news last night about IBC activities. They are rebuilding their studios at 37 Portland Place and expect to run three transmitters in France.[40]

Street (2006, 201) notes that Leonard Plugge was even, in 1947, attempting to buy a new transmitter in the USA and relaunch broadcasts from Normandy. Alas he found the French administration was adamant about its post-war broadcasting policy, and the possibility that Radio Normandy might once again broadcast became an unattainable goal. Instead, the IBC headquarters in Portland Place became an independent recording studio.

---

[39] Memo from Haley to L. Wellington, 31 December 1945. BBC WAC: R34/420, Home Services.

[40] BBC Memo from Chalmers to Collins, 23 July 1946. BBC WAC E2/365/2.

Radio Luxembourg did not suffer a similar fate. Haley reported in May 1946 that Radio Luxembourg was to recommence full commercial broadcasting on 1 July and added that this meant it was time to 'look at possibly amending the output of the Home Service on Sundays and perhaps extending the hours of the Light Programme.'[41] As Street (ibid, 197) remarks, the fact that Radio Luxembourg was captured twice, by the Germans in 1940 and by the Americans in 1944, without damage to its broadcasting or transmission equipment shows how determined both the German and allied forces were to keep the station available for propaganda purposes. When the war ended on 8 May 1945, the allies continued to control the station for two months[42] until it was handed back to its previous owners. Radio Luxembourg officially began broadcasting again in French on 12 November 1945, with the famous words: 'Bonjour le Monde, ici Radio Luxembourg.' When English-sponsored programmes began again on 1 July 1946 they did so under Stephen Williams as Director of English programmes, a position he had held previously at the station since 1933 (Williams 1987).

Radio Luxembourg became the sole English language competitor for the BBC in the immediate years following the war. It had been a familiar station before the war so found no difficulty in attracting British listeners again. The BBC of course was wary of its arrival on British shores after almost seven years and was concerned how 'the habit of listening to Radio Luxembourg, particularly along the South coast towns was growing rapidly.'[43] Also, on learning that the station was recommencing broadcasting, Haley was keen to examine areas of the BBC schedules which might need to be reappraised if they were likely to be targeted by Radio Luxembourg.[44] But, coupled with this was a typical elitist BBC attitude to sections of the audience. In one report, its author, Leslie Baily, states:

> I am a family man in his forties, a member of the English nonconformist middle class... and already had an antipathy towards sponsored radio... which is probably shared by many of my class.[45]

[41] BBC Programme Policy Meeting Minutes, 21 May 1946. BBC WAC R34/454/1.

[42] The station was called Radio Twelve Twelve after the US 12th army division which operated the broadcasts.

[43] Report by Kenneth Adam, Radio Luxembourg Developments July 1949. BBC WAC E1/1054.

[44] Haley to Programme Planning Meeting, 21 May 1946. BBC WAC: R34/420, Home Services.

[45] Radio Luxembourg: A Report on its Programmes, Leslie Baily 1949. BBC WAC E1/1048.

It would appear a misappreciation of the audience still prevailed at the BBC and that the years of monopoly during the war had, if anything, allowed such an attitude to fester. One thing that was clear however was that a policy of non-cooperation with competitors at any level remained intact. When the Director General of Radio Luxembourg wrote to his BBC counterpart asking if the corporation might consider taking part in a new programme looking at youth across different European countries, the answer was a definite 'no,' and in fact the letter has scrawled across it in red ink by an unknown hand 'we do not want anything to do with them.'[46]

Radio Luxembourg was to build on its success not only as an entertainer but also as an innovator. It secured *Opportunity Knocks* (1950) from the BBC[47] and broadcast it until it was transferred to television on the nascent ITV in 1956. Radio Luxembourg was also responsible for Britain's first *Top Twenty* chart show (1948), and a future innovation for this show came in 1952 when the chart began representing gramophone record sales rather than sheet music sales. Street (2006, 199) identifies four phases in Radio Luxembourg's subsequent long-term post-war re-emergence which are useful to cite at this point: firstly, the post-1946 'family' broadcasting which was very similar to its pre-war model; secondly, the post-1955 move to becoming a niche record-based station rather than entertainment-based which resulted from the arrival of competition in the form of commercial television which essentially provided a visual dimension to the type of output broadcast by Radio Luxembourg; thirdly, the post-1964 erosion of its position with the arrival of the pirate stations; and finally, the post-1967 decline after the arrival of BBC Radio 1. One could add to this a fifth phase, namely, British legislation which introduced legalised commercial radio in 1973 and which led to the ultimate demise of Radio Luxembourg in 1992.

THE REVIVAL OF TELEVISION

On the morning of 1 September 1939, Douglas Birkinshaw, the engineer in charge at Alexandra Palace, received a message at 10.00am stating that the BBC's television station would be closed by noon. The last item to be televised was a Mickey Mouse cartoon—there was no closing

---

[46] Letter from R.L. Peulvey, Directeur General Radio Luxembourg, 24 March 1947. BBC WAC E1/1054.

[47] The talent show began on the BBC Light Programme in February 1949.

announcement. Fears that television masts could be used as navigational aids to enemy aircraft meant the service was to be abruptly curtailed by political events and would remain closed for the duration of the war. The service resumed in June 1946, with the interrupted Mickey Mouse cartoon from 1939 its first broadcast. It steadily began to grow, aided by a new combined television and radio licence with the number of combined licences reaching 7.3 million by 1957, equalling the number of sound only licences for the first time.[48] During the summer of 1948 a BBC Survey on TV watching habits among 900 viewers found 91% switched in most evenings and remained viewing, although 16% said they had to make frequent adjustments to the set (Kynaston 2007, 214). Television still had some technical problems which meant radio maintained its position as the primary medium, and this trend would continue for radio. According to a BBC Audience Research newsletter from November 1950, the new series of *Take It From Here* was heard by 38% of the adult population, and 57% were listening to *Variety Cavalcade*. One in three were listening to *Educating Archie* by the end of its first series in the summer of 1950, and at 11.00am each weekday *Mrs Dale's Diary* was listened to by 13% of adults (Kynaston ibid, 583).

During the war the BBC had not devoted anywhere near as much discussion on the future of television as it had done for the future of radio, other than an occasional mention of the need to be prepared to take an early decision on 'the technical standards to be adopted for the resumption of a television service.'[49] This reflected a somewhat ambivalent approach. Haley continued to share Reith's dislike of television, still believing by 1949 that television was merely an extension of sound broadcasting. Crisell (2002, 80) argues that after the war this meant television resumed 'not so much with the expectation that it would develop as that it would need to be curbed,' and McKibbin (1998, 471) goes further by concluding that 'although television was predictably the wave of the future... the BBC had little interest in it'. So it appeared the support which had been engendered in 1939 may have dissipated after 1945; this may not be surprising however, considering the hugely important role radio had played over the course of the war.

[48] Report: The Future of Sound Broadcasting in the Domestic Services, January 1957. BBC WAC: R34/1021.
[49] Post-War Planning Meeting, 13 November 1942. BBC WAC: R34/578/1.

The 1949 Labour government appointed a committee to look into the state of broadcasting and the resulting Beveridge Report[50] was published in January 1951, recommending a continued BBC monopoly with regard to television, although among the appendices was a minority report by Selwyn Lloyd (MP Con) advocating competition with a service paid for by advertising. There also existed in Parliament a small group of Tory backbenchers who formed a commercial television lobby. It would appear the calls for competition in the supply of television were gaining momentum. Despite this, as the war years had been for the BBC and radio, the years from 1946 to 1955 represented a period of monopoly for the BBC and television. However, one also begins to notice a change in the fortunes of the two media as a direct result of being in competition with each other with BBC suggestions that television may have accounted for half the drop in radio listening figures in the period between 1948 and 1956.[51] This element of competition within the BBC culminated in the coronation of Queen Elizabeth II on 2 June 1953 which is widely regarded as the moment when television superseded radio as the most significant mass medium. The event prompted a boom in television set sales and marked the moment when more television sets than radio sets were manufactured (Crisell 2002, 81). Crisell goes on to say however that the coronation merely 'hastened a trend that already been happening,' and by 1955 viewing figures had begun to exceed listening figures for the first time (Crisell ibid).

What is certain is that television's ascendancy was established in the early years of the 1950s, but the BBC's monopoly was soon to be broken as commercially funded public service television broadcasting was introduced to Britain on 22 September 1955 following the passing of the 1954 Television Act. The establishment of ITV[52] in 1955 can be seen as:

> ... the end product of a battle between those with commercial and
> ideological interests in the expansion of television services and
> those who feared the impact of commercial forces on British cultural
> life and its main impetus came from television set manufacturers,
> advertisers and other commercial interests. (Johnson and Turnock 2005, 16)

[50] Report of the Broadcasting Committee 1949 (Beveridge Report), Cmd 8116, London HMSO.

[51] Report: The Future of Sound Broadcasting in the Domestic Services, January 1957. BBC WAC: R34/1021.

[52] Launched in 1955 as Independent Television under the auspices of the Independent Television Authority.

As Radio Luxembourg had arrived back as a threat on the radio scene, ITV now represented an equal threat on the television landscape. By 1955 both the BBC's monopolies in radio and television were under threat.

CONCLUSION

In many ways the war years and the post-war years represented the zenith of radio broadcasting in the UK, fulfilling a significant role as informer, educator and entertainer during the years of wartime hardship and post-war austerity, and while television began to encroach on radio's erstwhile hegemonic status, the latter continued to flourish during this period, and, for many, programmes such as *The Goons* (*1951*), *The Archers* (*1951*), *Under Milk Wood* (*1954*) and *Hancock's Half Hour* (*1954*) mark a heyday of British radio broadcasting.

Street (2002, 82) believes British radio was redefined more in the period 1939–1945 and there are a number of reasons for reaching such a conclusion. Certainly, the BBC changed into something that had little in common with the Reithian model it had been before the war. Also, circumstances put BBC Radio at the heart of the British nation and as an external broadcaster it had also built a reputation for unbiased news reporting. An absence of competition gave the corporation time to address its pre-war position and make changes which would ameliorate its place in the post-war era. This was a slow process however as the memory of the threat of pre-war competition did not signal any significant changes to its programming until the latter stages of the war instilled change through partnerships with American and Canadian broadcasters. It is interesting to speculate how the broadcasting map of Britain would have evolved had war not broken out. Street (ibid, 190) goes so far as to say that 'the war saved the BBC' by removing all pre-war competition at a stroke and by bolstering the BBC's position as the sole provider of information.

Although BBC Radio faced little radio competition in the immediate post-war era, with Radio Luxembourg being the only serious competition, the steadily increasing demand for television and the arrival of ITV in 1955[53] meant BBC Radio was to suffer from competition from within, as television was to become the favoured medium inside the BBC itself in terms of resource allocation, particularly after the coronation of Queen

[53] BBC Radio went head to head on ITV's opening night with the death of Grace Archer in The Archers.

Elizabeth II in 1953, when BBC Television coverage achieved a peak viewing audience of over 20 million, overtaking the radio audience of 12 million for the first time. Further events would consolidate television's rising success, such as the televising of the 1953 FA Cup Final (Johnes and Mellor 2006).

The period of the Second World War and the immediate post-war period can be characterised by a number of factors. Firstly, it removed all elements of competition, thus giving the BBC the chance to regain its position. Unfortunately the BBC did not grasp this opportunity as eagerly as one might think, a nonchalance persisted in its attitude to competition, which is summed up by a report which concluded 'the BBC has little to fear from Radio Luxembourg.'[54] Secondly, it saw the emergence of *niche broadcasting* or *narrowcasting*[55] through the BBC's development of three separate post-war radio services from a sole service which had existed before 1940. Thirdly, it witnessed the emergence of a new technology—television—which was not just an extension of an existing technology but a new technology and one which would begin to overtake radio in this period. Relations between the BBC and the commercial sector were non-existent during the war simply because there was no commercial sector. The only relationship that did exist was with American and Canadian forces radio, and while this was short-lived, it did furnish the BBC with a greater insight into the world of *light* listening and provided it with an impetus to aim in such a direction. After the war the BBC again became a self-sufficient player, and it can be said there was no element of a relationship with Radio Luxembourg at any level. Radio Luxembourg continued to perform the important role of the commercial companies which was to demonstrate creativity with regard to populist content, and the introduction of the Top Twenty Chart Show illustrates how good the BBC's opponents were at leading the way in this type of programming. As the 1950s came to an end and we moved into the 1960s, Radio Luxembourg would no longer be the only significant competitor. The country was soon to be bombarded by broadcasts from vessels in the North Sea.

[54] Radio Luxembourg: A Report on its Programmes, Leslie Baily, November 1949. BBC WAC: E1/1048.

[55] The dissemination of targeted content to a narrow audience, not to the broader public at large. A concept that would dictate many future radio stations.

REFERENCES

Baade, C. *Victory through Harmony: The BBC and Popular Music in World War II.* New York: Oxford University Press, 2012.

Baily, L. *BBC Scrapbooks, Volume 2* London: George Allen and Unwin, 1968.

Barnard, S. *On the Radio: Music Radio in Britain.* Milton Keynes: Open University Press, 1989.

Briggs, A. *The BBC: The First Fifty Years.* Oxford: Oxford University Press, 1985.

Briggs, A. *The History of Broadcasting in the United Kingdom. Volume 2: The Golden Age of Wireless.* Oxford: Oxford University Press, 1995a

Briggs, A. *The History of British Broadcasting in the United Kingdom. Volume 3: The War of Words 1939–1945.* Oxford: Oxford University Press, 1995b.

Briggs, A. *The History of Broadcasting in the United Kingdom. Volume 4: Sound and Vision.* Oxford: Oxford University Press, 1995c.

Bridson, D.G. *Prospero and Ariel.* London: Victor Gollancz Ltd, 1971.

Carpenter, H. *The Envy of the World: Fifty Years of the BBC Third Programme and Radio 3.* London: Weidenfeld and Nicholson, 1996.

Chignell, H. *Public Issue Radio: Talks, News and Current Affairs in the Twentieth Century.* Basingstoke: Palgrave Macmillan, 2011.

Cole, J. *Lord Haw-Haw—and William Joyce: The Full Story.* New York: Farrar, Straus and Giroux, 1964.

Crisell, A. *An Introductory History of British Broadcasting* 2nd edition. London: Routledge, 2002.

Curran, J. and Seaton, J. *Power Without Responsibility: The Press, Broadcasting and New Media in Britain.* London: Routledge, 2003.

Dutton, D. *Reputations: Neville Chamberlain.* London: Bloomsbury Academic, 2001.

Haley, W. "The Wartime Administration of the BBC." *Public Administration* 24, no. 2 (1946).

Hilmes, M. "British Quality, American Chaos: Historical Dualisms and What They Leave Out." *The Radio Journal: International Studies in Broadcast and Audio Media* 1, no. 1 (2003).

Hobsbawm, E. *Age of Extremes: The Short Twentieth Century 1914–1991.* London: Michael Joseph, 1994.

Johnes, M. and Mellor, G. "The 1953 FA Cup Final: Modernity and Tradition in British Culture." *Contemporary British History* 20, no. 2 (2006).

Johnson, C. and Turnock, R., eds. *Independent Television Over Fifty Years.* Maidenhead: Open University Press, 2005.

Kenny, M. *Germany Calling: A Biography of William Joyce, Lord Haw-Haw.* Dublin: New Island, 2003.

Kuhn, R. *The Media in France.* London: Routledge, 1995).

Kynaston, D. *Austerity Britain 1945–51.* London: Bloomsbury, 2007.

Lewis, P. "Remembering Radio." *The Radio Journal: International Studies in Broadcast and Audio Media* 11, no. 1 (2013).

Lewis, P. and Booth, J. *The Invisible Medium: Public, Commercial and Community Radio*. London: Macmillan, 1989.

McKibbin, R. *Classes and Cultures: England, 1918–1951*. Oxford: Oxford University Press, 1998.

Martin, B. "Postwar Austerity to Postmodern Carnival: Culture in Britain from 1945." In *The Great, The New and The British: Essays on Postwar Britain,* edited by Ribberink, A. and Righart, A. Hilversum: Verloren, 2000.

Morley, P. *'This is the American Forces Network': the Anglo-American Battle of the Airwaves in World War II*. Westport, CT: Praeger Publishers, 2001.

Nicholas, S. "The People's Radio: The BBC and its Audience, 1939–1945." In *'Millions Like Us'?: British Culture in the Second World War,* edited by Hayes, N. and Hill, J. Liverpool: Liverpool University Press, 1999.

Pawley, E. *BBC Engineering 1922–1972*. London: BBC Publications, 1972.

Plomley, R. *Days Seemed Longer*. London: Eyre Methuen, 1980.

Pronay, N. and Taylor, P. "'An Improper Use of Broadcasting...' The British Government and Clandestine Radio Propaganda Operations Against Germany During the Munich Crisis and After." *Journal of Contemporary History* 19, no. 3 (1984).

Reid, C. *Action Stations: A History of Broadcasting House*. London: Robson Books, 1987.

Street, S. *A Concise History of British Radio 1922–2002*. Tiverton: Kelly Publications, 2002.

Street, S. *Crossing the Ether: British Public Service Radio and Commercial Competition 1922–1945*. Eastleigh: John Libbey Publishing, 2006.

Williams, S. "Pioneering Commercial Radio the "D-I-Y" Way." *European Journal of Marketing* 21, no. 8 (1987).

CHAPTER 4

Commercial Onslaught: Commercial Television, Radio Luxembourg and the Pirates

From the mid-1950s commercial competition for BBC Radio was coming back into force. Some of it would resemble previous incarnations, some of it would use different tactics and some of it would come from a completely different media source. Commercial television would have an impact not just on BBC Television but also on BBC Radio. In fact, one could argue that while the BBC's own television output would have repercussions for its radio offering, it was commercial television which had a leading role in what can be described as the switchover from radio to television 'as the main household activity in the evening' (Hendy 2000, 125). Of course one could argue that the switchover took place at one point in time, namely, the BBC's broadcast of the coronation of Queen Elizabeth II in 1953, but it is also apt to argue that this switchover happened over a longer period rather than with a single event and that the impact of commercial television helped to ensure it was an irreversible phenomenon.

In looking at the BBC's radio competitors, Radio Luxembourg remained for some time the only real opposition. Commercial television had its own impact on Radio Luxembourg too and the latter had to reinvent itself in the late 1950s in order to survive, something it did quite successfully and found itself again representing the major radio threat to the BBC. It was however the arrival of the pirate radio stations in the early 1960s which were to really force the BBC—and eventually the government—to take action. Along with Radio Luxembourg, the pirates captured large swathes of the popular music audience, enough to startle the

© The Author(s) 2018
JP Devlin, *From Analogue to Digital Radio*,
https://doi.org/10.1007/978-3-319-93070-1_4

BBC. Also at this time, the campaign for a wider choice in radio offerings went through a renaissance as pressure groups demanded space on the wavelengths for non-BBC radio stations and those which could operate in a totally legal fashion. It is no mere coincidence that Briggs gave the simple title *Competition* for Volume Five of his seminal *History of Broadcasting in the United Kingdom* (Briggs 1995) which covers a similar period, as these years were marked by the arrival of competition on numerous fronts and where, yet again, radio competitors took advantage of the BBC's slow reaction to audience needs. Competition took many forms, and the distinct and separate nature of the BBC and its competitors simply reinforced a climate of estrangement which typifies the period.

COMMERCIAL TELEVISION

The opening night of Independent Television (ITV), the UK's commercial public service television network, on 22 September 1955, marked the end of a 'protracted and bitter' (Briggs 1995, 3) debate on whether or not to break the BBC's television monopoly. In addition to the beginning of a further consolidation of television as the primary mass medium, it also heralded another threat to the position of BBC Radio on the media landscape. Television as a medium under a BBC monopoly had been growing steadily since the service relaunched in 1946 and began to encroach on radio's status, it then began to quickly find its place in British living rooms due to the actions of both the BBC and ITV. Between 1946 and 1962, annual issues of the combined radio and television licence grew from 15,000 to 12 million,[1] marking the zenith for the radio licence which eventually declined to the point where the separate radio licence was finally dropped in February 1971. By 1962 the BBC was estimating 22 million people in the UK were regularly watching television whether BBC or ITV.[2] But while both entities were creating this new television world, the launch of commercial television marked a twofold threat to the BBC. It threatened the BBC's existing television service, but it also created a new threat to its radio service which had already been suffering under the corporation's own tentative steps into television broadcasting, which had resulted in television accounting for the major proportion of BBC expenditure after 1958. It was around this time in the late 1950s which marks

[1] British Broadcasting 1922–1962, BBC Press Office, 14 November 1962.
[2] ibid.

what Winston describes as the true start of 'television diffusion' where it could be said television had finally arrived not only in living rooms but also on factory floors:

> Everywhere, radio manufacturers and producing entities switched to television. There were no casualties. There were few new faces. (Winston 1998, 125)

It would appear that just as the BBC had established television and indeed had transformed itself into a bi-media entity, did the commercial television lobby and the industry (both manufacturers and retailers) simply diversify and carry on. The BBC began, at this point, gathering information on its competitors' plans, much as it had done with the IBC in the 1930s. So it was able to work out that Granada Television[3] was in 1955 negotiating on a four acre site to greatly extend its operations near Manchester city centre.[4]

ITV had however actually got off to an uncertain start. The franchise companies initially struggled to attract advertising revenue and there was a genuine concern that 'the commercial enterprise might prove to be a failure' (Turnock 2007, 165). To receive the new service, viewers needed to fit a special aerial and either buy a new television set or have their existing set adapted. In the first franchise areas only 33% of households took these measures (Crisell 2002, 93). For the first few months of commercial television, it could only be seen in the London area but then began to spread quite quickly.[5] By the second half of 1956, only a year after launch, the tide began to turn. By September 1955 188,000 homes were equipped for ITV, by September 1956 this had risen to 1.5 million, by 1958 it was 5.25 million and by 1960 over 5 million with estimates that over autumn and winter 1956/1957 viewers were increasing by 50,000 per week (Crisell ibid). Despite this early success the commercial companies were initially making a loss. Associated-Rediffusion[6] had lost more than £2 million by the end of 1955, while Associated Television (ATV)[7] lost nearly £3 million by the end of the first year of trading. Between 1955 and 1957, Associated-Rediffusion lost a further £3 million yet only two

[3] Based in Manchester, it was awarded the ITA franchise for the North of England in 1954.
[4] The Competitor. BBC Document, 1955. BBC WAC: R34/1177/1.
[5] The regional franchises were complete by 1962.
[6] Holder of the London weekday franchise.
[7] Holder of the London weekend franchise.

years later it began to turn annual profits of £5 million (Crisell ibid). Thanks to the change of fortune most of the commercial companies moved to profit in the second year and lavish returns by the third, independent television had in effect become 'a goldmine' according to Briggs (1995, 9). The few well organised entrepreneurs who recognised the potential of the new medium for making money were suitably rewarded. Roy Thomson of Scottish Television candidly stated, 'running a commercial television station is like having a licence to print money' (Black 1972, 55). From the outset, Granada set itself the objective of providing high quality popular programmes that would surpass even the BBC's threshold of quality and standards. Plowright (2003) claims:

> Granada television was the most precocious of the independent television companies formed in the 1950s. It was also the most innovative, self-opinionated, insufferably arrogant for some of its competitors and defiant of authority if it tried to obstruct the transmission of programmes Granada considered to be in the public interest. It was swashbuckling and successful and it was irresistible for the first generation of commercial television programme makers keen to challenge the monopoly of the BBC.

The stark effect commercial television had on the BBC was revealed less than two years after its arrival by the corporation's own research which showed that by July 1957, in homes with a choice of channels, three adults watched ITV for every two who watched the BBC.[8] Crisell (2002, 10) says the problem was not now with BBC programmes but with scheduling, namely, broadcasting heavy programmes at peak competitive times; thus the imperative to evaluate scheduling was now greater than ever, as while the BBC retained moral and cultural superiority, commercial television was winning on audience share. Commercial opposition's ability to target audience needs satisfactorily was once again the emergent theme.

The low point as far as the BBC's audience was concerned came in the third quarter of 1957 when ITV, on the BBC's own calculations, gained a 72% share of the viewing public wherever there was a choice. It was in light of such statistics that BBC programming policies changed with a well-planned counter-attack. New programmes designed to appeal to a young audience were introduced such as *Six-Five Special*.[9] But as the

[8] Annual Report & Accounts of the BBC, 1958, p. 8.
[9] The BBC's first attempt at a television *Rock 'n' Roll* programme, launched in February 1957.

governors insisted in their annual report to parliament in 1957, success in a battle for audience ratings should not be the sole aim of the BBC:

> A full and fair provision must be made for the exploitation of what is best and most suited to the medium without sole regard to size of audience.[10]

During the first few months of commercial television's existence a trend began where the same number of television sets were being sold as radio sets and this equal trend continued for the next year.[11] Briggs (1995, 30) argues that the years between 1957 and 1959 represent the turning point when television overtook radio within the mindset of the BBC in practical terms as expenditure in the period 1955/1956 on radio amounted to £10,930,584, while on television it was £7,033,044. By 1958/1959 this had gone up to £11,441,818 on radio and £13,988,812 on television. So it was at this point when television became the dominant medium for the corporation. Of course commercial television had its detractors. The leader of the opposition, Herbert Morrison, believed it to be a 'threat to the national economy' (Morrison 1953), while the Archbishop of Canterbury, Geoffrey Fisher, condemned the pouring of millions of pounds into establishing the enterprise.[12] But television had taken hold in BBC policy terms as it was recognised that:

> ... television has developed so successfully that we must soon begin to look upon it as playing the major part in the general task of broadcasting for the home audience.[13]

It is important to point out that the fall in radio listening which was occurring around the latter half of the 1950s[14] was not deemed within the BBC to be solely due to television but was also thought to be due to other factors such as 'changes in social habits and the growth of other distractions,' and it was believed within the BBC that these factors alone could

[10] Annual Report & Accounts of the BBC, 1958, 8.

[11] *The Economist*, 14 January 1956.

[12] BBC Document. Policy: Commercial Television, 28 October 1955. BBC WAC: R34/1177/1.

[13] BBC Annual Report, 1956/57, 5.

[14] For example, by the end of the 1950s evening time listening had collapsed to 20% of its 1950 audience.

account for the huge fall in listening.[15] The Board of Governors remained ebullient claiming sound radio would continue to have a 'vital role to play in an age of television broadcasting.'[16]

Director General Ian Jacob[17] told sound staff they still had an important role to play and that 'there was a real challenge backed by the needs of 9 million listeners.'[18] The BBC's internal response was to set up a committee to reappraise the role of sound radio within the corporation. Known as the Sound Co-ordinating Committee,[19] it first met in 1955 and its role was to examine the current state of sound broadcasting and make recommendations on its future shape and policy. The report produced by the committee in 1957 was called The Future of Sound Broadcasting in the Domestic Services[20] and suggested the corporation should seek to 'cater for the needs and tastes of its audiences without seeking to alter or improve them as it had in the past.' The report also noted the 'profound change of mood in the country since the end of the war, especially among the young.' One can accuse the BBC of simply going over the same ground. After all, notions of 'lightening' output and catering for wider tastes and providing more entertainment had been seen as the cure for the many ills the BBC had suffered in previous years, from the days of the commercial radio opposition of the late 1930s to the days of the Forces Programme during the war to the days of the arrival of commercial television, but this report seemed to finally take on board the social changes which were taking place and was beginning to take seriously the notion of *youth culture*, insisting that radio now 'needs to battle to regain ground.' This led to a re-ordering of the BBC's radio output, but would it be enough or was it merely tokenistic?

Jacob seems to have been aware that the only course of action for the BBC was to compete for audiences and his stance was fully supported by the governors. So inside the BBC there at least seemed to be an awareness that the corporation would now have to compete, although this does not

[15] BBC Report: The Trend of Listening Since the War. R. Silvey, 4 January 1957. BBC WAC: R34/1022/3.

[16] BBC Board of Governors minutes, 25 September 1957. BBC WAC: T16/310/1.

[17] BBC Director General, 1952–1959.

[18] *Ariel*, Winter 1955.

[19] Chaired by Frank Gillard who would later play an important role in the launch of BBC Local Radio.

[20] BBC Report: The Future of Sound Broadcasting in the Domestic Services, January 1957. BBC WAC: R34/1021.

appear to have been an aggressive policy. Perhaps the fact that the press were initially fearful of an independent television service meant that the BBC thought such a position taken by the press might mean that the job could be done for them and an aggressive competitive policy might be unnecessary. BBC Director of Television Gerald Beadle[21] (1963, 76) believes that if the BBC had demonstrated its 'full competitive strength' during the first two years of competition by providing entertainment between 7.00pm and 10.30pm, it could have put ITV out of business. It would appear that the old chestnut of mass entertainment was still somewhat at variance with BBC policy however. Briggs (1995, 4) claims the next Director General, Hugh Greene,[22] disliked the constant striving to reach the largest possible audience for everything, while Beadle (1963, 77) goes on to say that such a policy of:

> ... lowering the proportion of intelligent programmes below the level of one's competitor would have opened up a vast wasteland from which it would have been impossible to recover.

The BBC appeared to be almost pretending competition did not exist and in fact the word *competition* was mentioned only twice in the *BBC Handbook* for 1957[23] and not at all in 1958,[24] almost as if it had magically disappeared. In terms of cooperation between the BBC and ITV, this was virtually non-existent except in the area of children's television when in 1959 the BBC and the Independent Television Authority (ITA)[25] set up a joint committee to consider a recent study in this area.[26] In April 1959 the Chairman of Associated-Rediffusion, John Spencer Wills, raised with Ian Jacob the possibility of some degree of collaboration between the BBC and the main television companies with a view to avoiding programming clashes. Jacob's response was a paper which outlined the fact that there was 'no legal bar under the BBC charter to such a degree of cooperation' but he believed the companies would not be allowed to proceed due to 'economic pressure brought upon them by creating such a degree of

---

[21] Became BBC Director of Television in July 1956.
[22] BBC Director General 1960–1969.
[23] *BBC Handbook*, 1957. London, BBC Publications.
[24] *BBC Handbook*, 1958. London, BBC Publications.
[25] The regulatory body for commercial television 1954–1972.
[26] See Himmelweit (1958).

collaboration.'[27] For the first time one can determine that the BBC is not against cooperation per se, nor would it be going against the principles of the charter. One can conclude that there is clearly, for the BBC, no obstacle to creating the conditions for cooperation with competitors and that any such decision, even at any point in the future, would simply be one of preference. In fact, Jacob also concluded cooperation should be a 'relatively easy step for the BBC and it would be commercial companies who would find any degree of cooperation difficult.'[28]

One potential method of taking on ITV head-on was of course to launch another television channel. The BBC had been considering a second channel for some time, whether this was in an attempt to deflect from the commercial companies is difficult to say, but it was certainly something it had been promoting even before the arrival of ITV.[29] It was in fact the Pilkington Report[30] of 1 June 1962 which led to the setting up of a third television channel and which actually placed responsibility for the new channel in the hands of the BBC. Chignell (2011, 82) sums up the Pilkington Report as:

> A hearty slap on the back for the BBC as it was critical of
> standards at ITV and gave the BBC a second television channel.

Chignell (ibid) goes on to quote BBC Director General Hugh Carlton Greene's smug reaction:

> The Pilkington Report seemed to us at the time a gratifying vindication
> of all that the BBC had been trying to do.

Awarding the new service to the BBC appeared to consolidate notions of public service broadcasting, highlight the negative attitude to commercial television output and therefore set out the model of broadcasting favoured at governmental level. The new service would operate on the new high-definition 625-line ultra high frequency (UHF) system which

---

[27] Note by Director General, Relations with Independent Television, 9 April 1959. BBC WAC: R6/239/1.

[28] ibid.

[29] BBC Annual Report & Accounts, 1954/55, 7.

[30] Report of the Committee on Broadcasting 1960 (Pilkington Report). June 1962. London: HMSO.

meant that to receive the new BBC channel, viewers would need a new set or adapt existing sets. Turnock (2007, 32) believes the setting up of BBC 2[31] marked an:

> ... uneasy conjunction of broadcasting policy, political debate, technological change and rapid expansion of the television industry.

Securing the third national television channel was good news for the BBC, it gave weight to the corporation's own view of being the nation's moral guardian in the face of low-brow entertainment and indeed having secured the third channel—and subsequently launching it as a channel catering for the type of programming for which the BBC had become renowned (educational and minority interest content)—this actually consolidated the enduring stance that the BBC's key role was to maintain a high level of responsibility in its public service remit. This was highlighted by Director General Hugh Greene speaking about the BBC as a public service broadcaster at the Manchester Luncheon Club on 9 November 1960:

> I have noticed a tendency... by some ITA spokesmen to make out that there is no real distinction between the BBC and commercial television... it is flattering that commercial broadcasting should wish to come under the public service umbrella but that umbrella belongs to us and there is no room under it for commercial broadcasting.[32]

The BBC appeared far from embattled by the introduction of ITV and clung steadfastly to the notion that it had a very specific role to play with a strong public service broadcasting remit which became almost further enshrined as a result of the findings of the Pilkington report.[33] However, this was coupled with a realisation that television would become a more powerful medium. Speaking in 1957, Director of Television, Gerald Beadle, predicted two thirds to three quarters of UK households would have television by 1960 and at a later point there would come a time when

---

[31] BBC 2 launched on 20 April 1964.

[32] The BBC as a Public Service. Speeches by H Carleton Greene, Director General of the BBC. *BBC Publications*, December 1960.

[33] In addition to a second BBC television channel, the report also makes recommendations on extended radio hours, the introduction of colour television and local radio broadcasting.

every household would have a multi-channel receiver.[34] By the early 1960s the BBC felt secure in its position in the British television arena and without having to change its attitude to audience needs. For radio, such a lax attitude would be challenged by the continued threat from Radio Luxembourg and from other radio interests lurking on the horizon.

COMMERCIAL RADIO

Until March 1964 the only significant form of competition the BBC faced in terms of sound broadcasting came from Radio Luxembourg.[35] As mentioned in the previous chapter, Radio Luxembourg recommenced its English language broadcasts on 1 July 1946, having been shut down for the duration of the Second World War. On 2 July 1951 it switched to the 208 metres medium wave wavelength which was to become well known to younger listeners throughout Britain and indeed would remain so until the station finally closed on 30 December 1992.

Radio Luxembourg began its post-war renaissance by doing exactly what it had done before the war, mainly because this had been a very successful policy direction and one which the BBC didn't impinge upon. So, as before, Radio Luxembourg continued to be supplied by agency production houses in London, recording programmes which would be broadcast back to Britain from the Grand Duchy, much as they had been done in the 1930s.

Radio Luxembourg's popularity returned as quickly as it retook to the airwaves, even despite the fact that it did not start broadcasting until 7.00 in the evening each day. Eventually programmes like *Opportunity Knocks* (1950)[36] and *Take Your Pick* (1953) began to become immensely popular among an audience which enjoyed quiz shows with big prizes as well as talent shows—that is, core light entertainment programming. American studies into television in the early 1960s were suggesting that entertainment could be integrated with serious content as surveys of American audiences provided evidence of television being considered as 'primarily a source of relaxation and entertainment' (McGhee 1980, 187).

[34] Television in Britain. Gerald Beadle (Director of Television). Speech to the Radio Industries Club, 29 October 1957. BBC Publications.

[35] Pirate station Radio Caroline began broadcasting on 27 March 1964.

[36] *Opportunity Knocks* started as a radio show on the BBC Light Programme in 1949 before moving to Radio Luxembourg.

A British study around the same time however claimed it was 'difficult to get many viewers to respond positively to be both entertained and educated at the same time' and that a lack of channel choice was the only reason for respondents to describe serious programmes as favourites, but 'as lighter programmes were delivered following the arrival of commercial television, such heavier programmes were watched less frequently' (Himmelweit 1962). Radio Luxembourg seemed to have its finger on the light entertainment pulse and these theories could equally be applied to radio and indeed help explain why Radio Luxembourg became so successful once again by offering lighter content.

Commercial television had a tremendous impact on Radio Luxembourg too and it may be argued that this impact was greater than that felt by BBC Radio at the same time. This was due to the fact that ITV directly affected Radio Luxembourg's revenue stream. The ITV audience increased vis-à-vis the BBC in terms of the BBC/ITV ratio which went from 54:46 at the end of 1955 to 38:62 one year later with the BBC's share dropping to only a third in 1958 (Silvey 1974, 187), and this rapid growth was not lost on Radio Luxembourg's programme sponsors who decided it would be more profitable to take both their products and their programmes to ITV. Hence, many of Radio Luxembourg's successful shows were taken up by ITV and the sponsors simply went with them. The effect of this was to force Radio Luxembourg to expand its record show portfolio and switch to sponsorship from record companies such as Decca, Capitol and EMI as an alternative form of revenue which could be guaranteed as Radio Luxembourg, unlike the BBC, was not restricted by an agreement with Phonographic Performance Limited[37] over *needletime*[38] as it operated outside British jurisdiction.

Hand in hand with the increased prominence of record shows on Radio Luxembourg came a record-based *disc jockey* (DJ) culture with presenters who went on to become household names among the young, thus setting in place a music radio format which would eventually become the standard across much future radio. At the time, the BBC's response could hardly have been described as competing with the relaxed presentation style of Radio Luxembourg. Crisell (2002, 136) believes the overlap between the BBC's three radio networks was such that the networks were not

---

[37] UK music licencing company and performance rights organisation formed in 1934.

[38] Created by the Musicians' Union and Phonographic Performance Limited to restrict the amount of recorded music that could be transmitted by the BBC. Abandoned in 1988.

sufficiently distinctive to command listener loyalty. So, in search of a particular kind of output, a listener might have to scan the schedules of two or even all three networks. It would however be unfair to suggest that the BBC did nothing whatsoever at this point since the Light Programme introduced landmark music programmes such as *Pick of the Pops*[39] in October 1955 and *Saturday Club*[40] in March 1957. However, these two programmes apart, there was little to appeal to fans of contemporary popular music anywhere on BBC Radio in the late 1950s and early 1960s. BBC output was regarded as too serious and demanding by many who were turning to Radio Luxembourg and the BBC was 'losing touch with what was going on in British society' (Scannell 2009).

Encouraging casual, background listening, coupled with the likely effect of the transistor, which in the early 1960s heralded a new era of portable radio technology, meant the drift towards Luxembourg was strengthened even further. Once again Radio Luxembourg had taken the lead, this time catering for a young constituency interested in the booming musical genre of Rock and Roll. However, a BBC Audience research report from 1954 suggests it was unlikely that Radio Luxembourg's audiences were causing a reduction in listening to the BBC but rather that Radio Luxembourg had picked up an audience 'from amongst people who weren't previously finding broadcasts to their tastes in the BBC schedules.'[41] Figures showed on Sundays Radio Luxembourg had a larger audience than the Home Service and sometimes rivalled the Light Programme too. Also, the Luxembourg audience on Mondays to Fridays was comparable to the Home Service.[42] But despite these figures the BBC stance remained the same, that is, it was catering for certain standards in public service broadcasting which seemed to carry greater value to the corporation than mere audience levels.

As Radio Luxembourg represented the sole radio threat and indeed the only other major actor in the radio sphere, it would seem logical that other commercial interests would once again strive to capitalise on the market potential which had been demonstrated by the continued growth of

[39] Originally a chart show *Pick of the Pops* still exists today on BBC Radio 2 although in a different format to the original programme.

[40] A music programme aimed at teenagers originally called *Saturday Skiffle Club* and ran until 1969.

[41] BBC Audience Research Report, February 1954. BBC WAC: 970, Luxembourg 1937–1954.

[42] ibid.

commercial television. As Fletcher (2008, 28) notes, television's share of total advertising revenue soared in the late 1950s, grabbing 23.3% of total advertising revenue by 1960 and:

> ... commercial television's intrusion into people's homes changed the relationship of advertisers with the public, and of the public with advertisers.

Such fertile ground for television could surely be replicated on radio. But of course it was not simply the lucrative advertising revenue which attracted voices which began to call for a commercial radio sector in Britain. For many, it was time to break the BBC's monopoly and if it could happen for television then why not radio? The success of the pirate radio stations in the early 1960s encouraged a number of interest groups to come to the fore to lobby the government for commercial radio and the break-up of the BBC monopoly. One such group was the National Broadcasting Development Committee chaired by Conservative MP Sir Harmar Nicholls which began by putting forward suggestions to the Postmaster General for controlled experiments with local commercial radio stations.[43] This followed reports that the BBC had been pressing Postmaster General Reginald Bevins for the authority to set up its own local stations.[44] In 1964, John Whitney of Ross Radio Productions co-founded the Local Radio Association which agitated for legalised commercial radio in the UK. Also involved was John Gorst who had worked for Pye Ltd. Gorst, who later became a Conservative MP, was to be an important figure in lobbying Tory support for independent radio prior to the 1970 general election. In the same way that interest groups campaigning for commercial television had emerged from the Conservative Party, the case for commercial radio attracted Tories to 'the cause of anti-monopolistic radio' (Jarvis 2005, 151). A campaign for commercial radio was gaining momentum and had come a long way from that described by Taylor (1953):

> The outstanding feature of commercial radio broadcasting, for example that in the USA, is its monotony. There is such a multiplicity of programmes that their effect becomes diluted and there is little serious listening...

[43] *The Tribune*, 5 June 1964.
[44] *Sunday Times*, 31 May 1964.

the case against commercial radio is so overwhelming that it is difficult
to see how anyone could support a proposal for it.

There was a feeling that the huge success of the pirate stations had
encouraged the commercialist lobby and that all that might be needed
would be a go ahead from the government for the BBC monopoly to be
broken.[45] The Local Radio Association and a company called Commercial
Broadcasting Consultants which was established by Tony Cadman and
Hughie Green in 1966 now became the major players in the movement for
the breaking of the BBC radio monopoly.[46] Street (2002, 107) cites Gorst:

> One of our aims within the Local Radio Association was to meet the
> argument of the BBC that there were not enough medium wave
> frequencies to permit the creation of a commercial radio network.
> So we commissioned the Marconi Company to research a feasibility
> study and this of course proved that there were sufficient frequencies
> available.

The Local Radio Association proposal was for a local radio network which
might number 150 local radio stations. Unfortunately, the Local Radio
Association's plans came at a time when the Labour government under
Harold Wilson held a very anti-independent radio policy. Street (2006, 202)
says it remains to be discussed how post-war agitation for an alternative to
BBC public service radio eventually led to the removal of the British radio
monopoly. This monopoly would not be removed under statute until 1973
but had been breached by illegal operations firstly from Radio Luxembourg
and then from the pirate stations who, it may be argued, did more to change
the face of radio broadcasting in Britain than any pressure group.

THE PIRATES

Official policy on sound broadcasting in Britain was contained in paragraph 20 of the government White Paper of July 1962[47] which stated:

> The (Pilkington) committee take the view that no additional national
> sound broadcasting services are needed and here the Government is
> in agreement.

[45] *The Tribune*, 5 June 1964.
[46] *The Listener*, 7 January 1971.
[47] White Paper on Broadcasting, 4 July 1962, Cmnd 1770. London: HMSO.

Paragraph 21 set out support for the BBC's intention at this time to extend the hours of the Third Programme and the Light Programme which were the only significant changes to be made to the corporation's radio output. It was precisely this environment which would prove to be the fertile ground from which a new vigorous radio model would emerge. It was to become known as Pirate Radio and the term was used on a general level to denote any form of unlicensed broadcasting. Chignell (2009, 137), however, rightly suggests the term is mainly associated with the 'historically specific case of offshore radio in Britain in the 1960s' and indeed it is these broadcasts from various vessels located off the coast of mainland UK which typify the primary characteristics of pirate radio, namely, its unlicensed nature and its DJ-led popular music output. An exception to the rule prohibiting a breach of the BBC monopoly was Manx Radio on the Isle of Man. From 1964 it was Britain's first legal land-based commercial radio station through a special dispensation from the UK government.

The BBC had been aware of new illegal broadcasts for some time. For example, by 1961 the station The Voice of Nuclear Disarmament was using the sound channel of the Crystal Palace transmitter after the latter had closed down each evening.[48] Only a couple of years later such broadcasts were coming from ships at sea and bombarding the mainland with pop music. So, the BBC noted in October 1964 a new station, Radio Albatross, had been launched by Lincolnshire businessman Mr Robert Tidswell and was broadcasting to Yorkshire and Lincolnshire from a ship anchored in the Wash.[49] Also in December 1964, the BBC's Sound Broadcasting Committee was informed that an unidentified station had been heard on 267 metres coming from the Thames estuary transmitting a good signal across London.[50] But it was earlier that year on 29 March 1964 that what would become one of the most successful pirate stations began broadcasting when Radio Caroline[51] made its first broadcast from international waters five miles off Harwich. There followed an explosion of stations which changed the course of youth music radio culture in Britain by introducing a veritable smorgasbord of constant, American-style,

[48] Illegal Transmitter. BBC memo from Head of Engineering, 27 November 1961. BBC WAC: TI6/411.

[49] BBC Extract, 7 October 1964. BBC WAC: T16/411.

[50] BBC Extract, 8 December 1964. BBC WAC: T16/411.

[51] Other pirate stations would follow such as Radio London, Radio Essex, Radio Scotland and so on.

DJ-fronted, pop music radio of a style unheard of in the UK and flying in the face of the BBC's continued post-war policy which largely ignored this style of content.

Radio Caroline was begun by Irish businessman Ronan O'Rahilly who had been an agent for singer Georgie Fame. Fame's group The Blue Flames had not been taken up by the major record companies and O'Rahilly wanted to make effective use of him and became inspired by successful offshore stations that had existed off Europe and Scandinavia for some years, many of which came and went but notable survivors included Radio Syd off the Baltic coast of Sweden and Radio Veronica off Scheveningen in the Netherlands and Radio Mercur broadcasting to Denmark from a site off Copenhagen (Street 2002, 109). O'Rahilly wanted to get his artist's records played on the radio so he bought a vessel, named it Radio Caroline[52] and began broadcasting. It does now seem somewhat unusual to see that O'Rahilly's primary motivation was to circumvent the record companies' control over popular music broadcasting rather than following the aspirations of other activists of the time who wanted to see the end of the BBC monopoly and seek opportunities for private investors. In effect his crusade, which would have long-term implications for the British radio industry, appears to have been personal rather than a realisation of the dreams of those keen to expand the radio market.

Briggs (1995, 503) describes the conditions into which Radio Caroline was born in March 1964 by highlighting how the style and content of pop music had changed since *Six-Five Special*. The emphasis had shifted from solo performers like Cliff Richard and Billy Fury to groups such as The Beatles. The consumption of pop music had therefore altered over a period of six years and it was stations like Radio Caroline 'which had catered for the new direction of pop and provided the vehicle which could carry the new consumers' (Clark 2014). During the summer of 1963, a series of 11 programmes about The Beatles called *Pop Go The Beatles* was broadcast on the BBC Light Programme which may be seen as catering for the new audience, but this was providing programming in the typical BBC model when the audience for this kind of music demanded essentially the music itself, something which the BBC did not provide and

---

[52] O'Rahilly named the station after Caroline Kennedy, daughter of US President John F. Kennedy.

which was beginning to cause some concern in the organisation where it was noted that:

> Radio Caroline consisted of a repetition of top ten popular songs. The interest
> in these transmissions among teenagers gives rise to some concern.
> Legislation to stop the broadcasts can not be expected for some time so the corporation needs to combat the pirates.[53]

At this stage any perceived tactics for combating the new competition centred around curtailing publicity, so the Director General ruled no broadcast publicity should be given to Radio Caroline or any other pirate station as it had been noted Granada TV was preparing a documentary on the subject.[54]

BBC attitudes to pop music remained equivocal despite the setting up of a new Popular Music Department in July 1963 and the arrival of *Top of the Pops* on television on 1 January 1964 which became an immediate hit. But what this highlighted was that 'television seemed to be carrying pop music more successfully than radio' (Frith 2002). On the radio, traditional programmes like *Family Favourites* (1945) and *Housewives' Choice* (1946) were taken more seriously than *Saturday Club* (1957). The BBC did nevertheless defend its position at the time to critics who suggested the pirates were providing a service wanted by the public and one which the BBC was unwilling to provide, by highlighting the fact that the BBC had its hands effectively tied on the matter due to needletime restrictions since it could only broadcast records for 28 hours per week spread over the three networks.[55] After publication of the Pilkington Report, a determined attempt was made by the BBC to have these hours increased from 28 hours a week for domestic radio to 75 hours a week, but there was objection from the Musicians' Union which feared this would prejudice the employment of musicians. An article in the *Radio Times* in October 1966 entitled 'Why no continuous pop? The BBC explains'[56] criticised needletime restrictions and laid the blame at the hands of the big record companies. In the meantime the appeal of the pirates continued to rise with figures suggesting

[53] BBC Extract, 12 May 1964. BBC WAC: T16/411.
[54] BBC Extract, 21 April 1964. BBC WAC: T16/411.
[55] Pirate Broadcasting. BBC Background Note, May 1964. Illegal Transmitters. BBC WAC T16/411.
[56] *Radio Times*, 6 October 1966.

pirate radio audiences now reached two million. An opinion poll from June 1964 showed 74% of people living in range of Radio Caroline thought the government should not try to stop the station broadcasting.[57] By 1966 a poll into audience size for commercial stations concluded 45% of the UK population listened to either an offshore station or Radio Luxembourg each week (Street 2002, 109). In the end it was the Labour Postmaster General Tony Benn who engaged directly with the pirates in a battle of the airwaves by introducing the Marine, &c., Broadcasting (Offences) Bill which would eventually close down this branch of radio broadcasting.

Sterling (2004, 381) identifies four elements which were critical in the development of British pirate stations. Firstly, it was a reaction to the continued monopoly of domestic British radio by the BBC. Secondly, the territorial jurisdiction of Britain at the time was defined as ending three miles offshore which made broadcasting from sea feasible. Thirdly, there was an ongoing struggle in British broadcasting between the elitists and the popularisers with the BBC seen as bastion of the former, and fourthly, the unique social climate of the 1950s and 1960s, with younger people's reaction to austerity centred around music, created a new, vociferous audience group. Pirate radio set out to exploit what it considered to be the BBC's lack of enterprise in the broadcasting of pop music. It was also critical of Radio Luxembourg for its dependence on the bigger record companies who, when they plugged their own records, paid for the airtime they took over. Part of the appeal of the pirates was that they were deemed rebellious and this appealed to 'the young rebels of the 1960s' (Boyd 1986). Another factor which Sterling fails to mention was that changes in technology helped drive access to the pirates, particularly with the arrival of the transistor, an item of technology which not only improved accessibility but also was rebellious in itself as it represented a shift from communal to personal listening, or indeed the ultimate extension of what Lacey (2013, 113) describes as the 'privatization of listening.' The combination of a new technology of radio listening along with a new platform of radio listening had the effect of inspiring a new tranche of commercial opposition to the BBC and of leaving the BBC once again attempting to play catch-up to its competitors' advances.

It is of course a mistake to assume the Marine, &c., Broadcasting (Offences) Act 1967 marked the end of pirate radio in Britain; on the

[57] *The Daily Mail*, 3 June 1964.

contrary, it 'continues to the present day' (Street 2006, 204), but the movement created by the activism of the offshore stations made change inevitable, and within two weeks of the pirates coming off air some of the presenters were broadcasting again, this time on the BBC on Radio 1 as the BBC finally launched a service of pop music aimed at a youth audience and in a single fell swoop regained its radio monopoly thanks again to legislation—much as it had done in 1939.

BBC Radio

As early as 1964, *The Daily Mail* reported that the BBC was planning a new pop music service.[58] In June of that year the Postmaster General, Reginald Bevins, revealed to the House of Commons that he had discussed this plan with the BBC, the main stumbling block being the fact that the corporation needed permission to broadcast more records, something which it had not been successful in securing in order to fill the extended hours of the Light Programme. Despite the high quality of many of its radio programmes, the BBC was continuing to lose listeners during the early 1960s. Crisell (2002, 136) claims that only the statutory limit the government had imposed on television transmission hours prevented radio from being hit even harder as 'television, especially in the evening was the main cause of radio's woes.' An audience research enquiry carried out in November 1958[59] revealed virtually everyone in the country could receive sound broadcasting and two thirds of people could receive television as well. The 'sound only' public was mostly older and of a higher educational level than the 'sound and television' public which would appear to justify the BBC's continued highbrow approach to its sound content as it was, one could conclude, catering for its audience's exact needs. Such a conclusion is indeed acceptable but it ignores the increasing appeal of competing pop music radio which in only six years' time would remove the last remnants of the younger 'sound only' public.

Yet, still in the early part of the 1960s, the BBC was sticking to the position it had held for the previous four decades, namely as the bastion of public service broadcasting in Britain, even though it was aware that changes were happening in radio consumption. Director General Hugh

---

[58] *The Daily Mail*, 3 June 1964.
[59] The Public and the Programmes: A Report on an Audience Enquiry. BBC Publications, December 1959.

Greene speaking to American broadcasting executives laid down the BBC's position for the new era:

> The new age of broadcasting which lies before us… should stand in the service of truth and… in the service of artistic truth as well as hard factual truth.[60]

The first significant changes the BBC made to its radio portfolio since the post-war restructuring of 1946 came about in 1957. A working party was set up under Director of Sound Broadcasting, R.E.L Wellington, to examine the existing service and make recommendations on its future shape and policy. The report, published in January 1957,[61] concluded 'the effort to improve public taste had given the public indigestion and it had turned away.' So, time was ripe to modify current policy substantially. It should seek to cater for the needs and tastes of its audiences without seeking to alter and improve them as it had in the past. In particular:

> Entertainment should not be undervalued or regarded just as a stepping stone to serious things. Less attention should be placed on spoken word and more on those who look to radio for relaxation and diversion.[62]

The report also noted the profound change of mood in the country since the end of the Second World War particularly amongst the young. It concluded that the three existing BBC radio services were still necessary but that any competition between them should cease. The report's recommended changes were accepted at the Board of Management meeting on 25 January 1957. The Board subsequently decided on a significant re-ordering for the Light Programme, the Home Service and the Third Programme, with proposals for the Third Programme coming under the greater degree of scrutiny.

Wellington noted the changes that had developed since 1946 on the media landscape, particularly the great strides made by television, both BBC and commercial. Interestingly, he makes no mention of Radio Luxembourg which would have been the main radio competitor at this

---

[60] British Broadcasting 1922-1962, BBC Press Office, 14 November 1962.
[61] BBC Report: The Future of Sound Broadcasting in the Domestic Services, January 1957. BBC WAC R34/1021.
[62] ibid.

time, illustrating perhaps a lack of continued importance which the BBC attached to Radio Luxembourg and how far its position of threat had diminished since the late 1930s. Wellington also noted audiences for television had gone up, while audiences for radio had gone down, and his response was to plan a reorganisation 'less drastic than that of 1946 and one which would represent a remodelling of programmes rather than programme services.'[63] The new pattern of programmes was scheduled to begin on 29 September 1957 and would include an extended Light Programme carrying a richer choice of entertainment; a Home Service consisting of more general news, music, drama, talks, discussions and comment; a partial merger of the Light Programme and the Home Service; and the main change which was the creation of two specialised programmes using the existing Third Programme frequencies, the Third Programme itself between 8.00pm and 11.00pm, and a new Network Three at the earlier time of 6.00–8.00pm.[64] Network Three[65] would accommodate a number of spoken word programmes that would be displaced from the Home Service and the Light Programme.

Criticism of the BBC's 1957 changes centred largely on the plans for splitting the Third Programme and was led by the Third Programme Defence Society,[66] a body whose membership included T.S. Eliot, Laurence Olivier, Michael Tippett, Ralph Vaughan Williams and Bertrand Russell. The society was concerned at what it perceived as the threat to the highbrow content only available on the Third Programme. However, even despite the fact that Wellington had acknowledged a new radio constituency in the form of youth, this constituency was not addressed in the 1957 changes except for a provision of a few extra hours of programming on the Light Programme.[67] There remained an area, as the 1950s progressed, which found the BBC wanting in the eyes of the young, as it had in the 1930s, in the field of popular music particularly after 1956 when Rock and Roll erupted in the USA. The BBC failed to acknowledge the new form of

---

[63] ibid.

[64] The BBC's Plans for Sound Broadcasting. R.E.L. Wellington (BBC Director of Sound Broadcasting). *Manchester Guardian*, 28 August 1957.

[65] For more on Network Three, see Carpenter (1996, 181).

[66] Devoted to maintaining the traditional high status of British Broadcasting and later known as the Sound Broadcasting Society (See Carpenter 1996, 171).

[67] BBC Report: The Future of Sound Broadcasting in the Domestic Services, Jan 1957. BBC WAC R34/1021.

music which was taking the youth movement by storm or even acknowledge those who were actively following it.

The next major analysis of BBC Radio came in 1964 when the new
Director of Sound Broadcasting, Frank Gillard, argued that 'if television
was to throttle radio then sound broadcasting would be dead by now'
(Gillard 1964). Instead, it was still thriving in his opinion. Gillard chose
his research carefully claiming that figures showed only 4% of the population declared they did not listen to the radio at all or had no radio sets, and
estimating that over the previous six years, the average daily radio audience had actually been rising steadily, with up to 28 million people listening at some point every day and that they were using their radios for
2.5 hours every day; thus he was able to conclude more people were in
effect listening and listening for longer. Gillard's bright picture also estimated that in 1964 radio's average daily audience was only one fifth below
that of television and that 2.5 million new radio receivers were sold in
Britain in 1962. Gillard did acknowledge that there were some countries
where radio was becoming extinct or becoming no more than an amplified
jukebox (in a nod to the pirates), but the thrust of his argument was that
radio was certainly changing and that there was an increasing readiness
amongst the audience to try listening to new and avant-garde works of
music, literature and drama and that 'BBC radio should give a lead in
these things' (Gillard ibid). It is odd that Gillard does not mention pop
music here when it was exactly this format which was impacting on radio.
Equally, in talking of competition he describes it in the form of 'television
services, record players, tape recorders, home movies' without any mention of the pirates nor the bodies calling for a commercial radio service in
the UK which suggests, like the BBC's attitude to Radio Luxembourg, the
idea of direct radio competition seemed unimportant. Gillard goes on to
outline his plans for radio which include the introduction of an additional
local radio service over large parts of Britain, as recommended in the
Pilkington Report.[68] He also sought to extend the Light Programme and
the Third Programme and aimed to improve the music service but with
the caveat that this would not be 'the thin edge of a wedge leading to a
gramophone record takeover' (Gillard ibid).

The BBC's position with regard to commercial competition in the
1950s had been bolstered by the Beveridge Report.[69] As Taylor (1953,

---

[68] This would lead to the launch of BBC Local Radio in 1967.

[69] Report of the Broadcasting Committee 1949 (Beveridge Report), Cmd 8116, London
HMSO.

357) notes, the Beveridge Committee concluded that one fundamental objection condemned commercial radio:

> The commentator, news reporter, entertainer etc ceases to be doing the job for his own sake. His purpose is not to inform or educate but to gather an audience for a sales talk... The case against commercial radio sponsoring is so overwhelming that it is difficult to see how anyone could support a proposal for commercial radio.

However by May 1964 the BBC noted that Radio Caroline programmes consisted of a repetition of top ten pop songs and the interest among teenagers gave rise to some concern particularly as legislation to stop the broadcasts could not be expected for some time.[70] Only one month later in June 1964, the idea of a BBC popular music service was first mooted as Postmaster General Reginald Bevins made his revelation in the House of Commons. At this time the BBC had begun pressing Bevins for authority to set up local stations.[71] Deriding the commercial sector's non-stop pop music and advertisements, the BBC wanted to offer a community service which would be educational as well as providing light entertainment and pop music. The reaction of the commercial pressure groups to this was that they could provide the same, but there was an actual public demand for non-stop pop music rather than 'some hybrid offering as proposed by the BBC.'[72] The battle with the pirate stations ultimately became a political one when Bevins' successor as Postmaster General, Tony Benn, took on the illegality of these stations in a process that culminated in the Marine, &c., Broadcasting (Offences) Act 1967 which came into force on the night of the 14/15 August, outlawing 'advertising on or supplying an unlicensed offshore radio station from the UK,' thus silencing the pirate stations, although Radio Caroline continued broadcasting as Radio Caroline International for some considerable time thereafter.[73]

Hand in hand with Benn's determination to put an end to the pirates was his equal determination to introduce a national programme of 'light'

---

[70] BBC Extract, 12 May 1964. BBC WAC: T16/411.

[71] The Future of Sound Broadcasting: Local Broadcasting, 25 November 1960. BBC WAC: R1/96/9.

[72] *The Sunday Times,* 31 May 1964.

[73] The end of the offshore Radio Caroline came when the Broadcasting Act 1990 and a storm that caused its staff temporarily to abandon the ship, caused the station to come ashore in 1991.

music, although not necessarily from the BBC—he even considered a service run by the Post Office called 'POP' (Benn 1996, 125). This position changed and Benn proceeded to promote the BBC as the viable provider of the alternative service (Benn ibid, 151), so by late 1966 the campaign to stop the pirates with a concomitant campaign to launch a pop music service had taken off. Only a few weeks after legislation stopped the pirates, the BBC launched its pop music service called Radio 1 on 30 September 1967 along with Radio 2, Radio 3 and Radio 4 in what was the most significant restructuring of its radio output since 1946.

CONCLUSION

This period is characterised by a number of salient features which mark it as different to the preceding period in radio history and particularly in relation to the BBC's position and the competitive threat it faced. Television became the dominant medium in this period, superseding radio to a very significant degree. By 1958 almost half the British population was regularly watching television on a Sunday night with 45% of working class viewers tuning in for an average of four hours a week, and this trend was matched by a desertion from BBC Radio evening programmes (Finch 2003, 3).

Focussing solely on radio one can clearly see parallels between the 1930s and the 1960s regarding competitive pressure on the BBC from commercial sources, but added to this was entrepreneurial and political pressure from those striving to break the BBC's monopoly. This had been a successful strategy for those who introduced commercial television and now it was hoped radio could deliver similar rewards. Defenders of the BBC of course maintained that commercial broadcasting existed solely to sell goods, while public service broadcasting aimed to serve the public. However, it can be argued that ITV, commercially funded by advertising revenue, both reflected and promoted the burgeoning consumer culture of the 1950s and was symptomatic of a 'rapidly changing social climate' (Turnock 2007, 198). Hopkins (1963, 403) notes the 'disquieting effect the arrival of commercial television had on some sectors of the establishment' and how television became the 'scapegoat for the general ills of society.' Television was situated at this time amid the changes that rocked British society in general. As (Finch 2003, 4) notes:

> Caught up in the consumer boom of the 1950s and the changing lifestyles of people eager to modernise not only their homes but also their social aspirations, British television grew up and came of age.

Another important social change of the time was the arrival of Rock and Roll and the emergence of a radically new popular music era. Kynaston (2009, 268) believes Friday 14 November 1952 marked the start of the modern British record industry as the *New Musical Express* (*NME*) of that date stated:

> For the first time in the history of the British popular music business an authentic weekly survey of the best selling *pop* records has been devised and instituted.

The new pop music scene and radio developed a symbiotic relationship. Both thrived off each other for a certain—youthful—section of the population and it was no surprise that from this pirate radio would be born. Street (2006, 202) claims the growth of the pirates between 1964 and 1967 mirrored the development of commercial radio in relation to the BBC in the 1930s and this is quite true, but the main difference between the 1930s and the 1960s was that radio was now being driven by a musically radicalised youth culture coupled with technical developments in the shape of the transistor which would have major significance for radio listening at the time and indeed on how radio would be consumed over the next 20 years. Commercial enterprise, yet again, seemed to be able to detect audience trends and attempt to match them only for the BBC to eventually, albeit slowly, attempt to play catch-up. It is important to recognise that while the BBC maintained its core values of public service broadcasting, its competitors continued to shape an alternative form of radio broadcasting and Radio Luxembourg's decision to pursue DJ-led programmes consisting of playing commercial records is testament to this.

Mendelsohn (1966) in his theory of mass entertainment blames elite media critics for fostering misconceptions about television's entertainment role, arguing that such elites were essentially protecting their own interests and refusing to accept television entertainment because it attracted people away from the type of output such elites were trying to promote. He argues that audiences instead wanted the type of escapism television could offer and that this was in effect an important social role. These arguments can apply to the history of BBC Radio up to this point and particularly with reference to music radio. This period in radio history is characterised by competition from another medium, greater competition within the same medium, increased pressure on the BBC radio monopoly and huge social and cultural change. The BBC adhered to its core elite principles while its competitors continued with their, what only be described as, anti-elite objectives. The BBC did take some minor steps to address its precarious position, but from 1967 it would make broad changes to its national and local radio presence, all just ahead of a new form of competition in the form of legalised commercial radio.

REFERENCES

Beadle, G. *Television: A Critical Review.* London: Allen and Unwin, 1963.
Benn, T. *The Benn Diaries: 1940–1990.* London: Arrow Books, 1996.
Black, P. *The Mirror in the Corner: People's Television.* London: Hutchinson and Co, 1972.
Boyd, D. "Pirate Radio in Britain: A Programming Alternative." *Journal of Communication* 36, no. 2 (1986).
Briggs, A. *The History of Broadcasting in the United Kingdom. Volume 5: Competition.* Oxford: Oxford University Press, 1995.
Carpenter, H. *The Envy of the World: Fifty Years of the BBC Third Programme and Radio 3.* London: Weidenfeld and Nicholson, 1996.
Chignell, H. *Key Concepts in Radio Studies.* London: Sage, 2009.
Chignell, H. *Public Issue Radio: Talks, News and Current Affairs in the Twentieth Century.* Basingstoke: Palgrave Macmillan, 2011.
Clark, R. *Radio Caroline: The True Story of the Boat that Rocked.* Stroud: The History Press, 2014.
Crisell, A. *An Introductory History of British Broadcasting* 2nd edition. London: Routledge, 2002.
Finch, J. *Granada Television: The First Generation.* Manchester: Manchester University Press, 2003.
Fletcher, W. *Powers of Persuasion: The Inside Story of British Advertising.* Oxford: Oxford University Press, 2008.
Frith, S. "Look! Hear! The Uneasy Relationship of Music and Television." *Popular Music* 21, no. 3 (2002).
Gillard, F. *Sound Radio in the Television Age.* BBC Lunch-time Lectures, Second Series, No.6. London: BBC Publications, 1964.
Hendy, D. *Radio in the Global Age.* London: Polity Press, 2000.
Himmelweit, H. *Television and the Child: An Empirical Study of the Effect of Television on the Young.* Oxford: Oxford University Press, 1958.
Himmelweit, H. "A Theoretical Framework for the Consideration of the Effects of Television: A British Report." *Journal of Social Issues* 18, no. 2 (1962).
Hopkins, H. *The New look: A Social History of the Forties and Fifties in Britain.* London: Secker and Warburg, 1963.
Jarvis, M. *Conservative Governments, Morality and Social Change in Affluent Britain, 1957–64.* Manchester: Manchester University Press, 2005.

Kynaston, D. *Family Britain 1951–57*. London: Bloomsbury, 2009.

Lacey, K. *Listening Publics: The Politics and Experience of Audiences in the Media Age*. Cambridge: Polity Press, 2013.

McGhee, P. "Toward the Integration of Entertainment and Educational Functions of Television: the Role of Humor." In *The Entertainment Functions of Television*, edited by Tannenbaum, P. Hillsdale, NJ: Lawrence Erlbaum Associates, 1980.

Mendelsohn, H. *Mass Entertainment*. New Haven: College and University Press Services, 1966.

Morrison, H. "Commercial Television: The Argument Explained." *Political Quarterly* 24, no. 4 (1953).

Plowright, D. "Granada, 1957–92." In *Granada Television: The First Generation*, edited by Finch, J. Manchester: Manchester University Press, 2003.

Scannell, P. "What is Radio For?" *The Radio Journal: International Studies in Broadcast and Audio Media* 7, no. 1 (2009).

Silvey, R. *Who's Listening? The Story of BBC Audience Research*. London: George Allen and Unwin Ltd, 1974.

Sterling, C., ed. *Encyclopedia of Radio*. London: Taylor & Francis, 2004.

Street, S. *A Concise History of British Radio 1922—2002*. Tiverton: Kelly Publications, 2002.

Street, S. *Crossing the Ether: British Public Service Radio and Commercial Competition 1922–1945*. Eastleigh: John Libbey Publishing, 2006.

Taylor, S. "The BBC or Commercial Radio." *The Political Quarterly* 24, no 4 (1953).

Turnock, R. *Television and Consumer Culture: Britain and the Transformation of Modernity*. London: I.B. Tauris, 2007.

Winston, B. *Media, Technology and Society. A History: From the Telegraph to the Internet*. Abington: Routledge, 1998.

CHAPTER 5

A Level Playing Field: The BBC and Independent Radio

From the late 1960s through to the late 1980s a number of changes would radically alter the UK radio landscape. These changes came in the form of a number of factors which would become even more prominent in later models of broadcasting. They include a wider and more tailored listening experience, the emergence of narrowcasting as a template and increased localism.

The BBC initiated a reordering and rebranding of its existing national radio portfolio by the end of the 1960s, thus leading to more targeted programming and the provision of services to deliver this content. With the demise of the pirates and hence a diminution of rivalry to the BBC, renewed calls for a form of commercial radio did not fall on deaf ears. Now was the time for commercial competition to gain a foothold in the UK radio market and in a wholly legalised form. The form that this new commercial rival took was unique to the UK and indeed led to it being referred to as *independent* radio rather than *commercial* radio. Some competitive elements persisted of course, including the indefatigable Radio Luxembourg and an eventual re-emergence of pirate stations in a different form.

The BBC also introduced a local radio service delivered at the micro level, thus marking a new direction from the macro level, a trend which would continue to characterise radio in general for the next number of years. With the introduction of licensed, local, commercial broadcasting, the BBC now found itself in competition with other radio entities, but this time they were entities competing on a level playing field

© The Author(s) 2018
JP Devlin, *From Analogue to Digital Radio*,
https://doi.org/10.1007/978-3-319-93070-1_5

constructed on the soil of statute. For the commercial sector, it was their time to prove that what they had to offer was appealing to the audience, that they could compete with the BBC and, most importantly, that they could make a profit in order to ensure their continued existence.

BBC RADIO: TWIN SERVICES—NATIONAL AND LOCAL

Two very important changes of direction took place at BBC Radio; these took place at a national level, namely, to the BBC's existing radio portfolio, and also at a local level where the BBC embarked upon the road of extensive localised broadcasting. It also saw the publication of a detailed plan for network and local radio, *Broadcasting in the Seventies*,[1] which arguably became one of the most important documents ever produced by the corporation in regard to its radio services and which Briggs (1995, Vol 5, xxi) describes as 'one of the most controversial documents in BBC history.' Changes to BBC Radio at this time were driven largely by competition—the experience of the powerful illegal competition of the 1960s as well as a fear of imminent legalised competition which appeared inevitable and therefore required a response on both a national and local level.

BBC National Radio

By the latter half of the 1960s the future of the offshore pirate stations, which had become the BBC's main radio competitor, was beginning to come to an end, and in 1966 a government White Paper on broadcasting[2] had given the BBC permission to open its own pop music channel. The Marine, &c., Broadcasting (Offences) Act became law on 14 August 1967, achieving a long-term government aim of 'outlawing broadcasts emanating from outside territorial boundaries' (Peters 2011) and a little over a month later, on 30 September 1967, Radio 1 launched on 247 metres medium wave. Radio 1 was only one in a new package of radio offerings from the BBC which now came into effect. In reality, Radio 1 was the only actual new station in this package and it was specifically created to offer a popular music service and crucially to offer such a service tailored to match the tone and energy that had been the successful

---

[1] Broadcasting in the Seventies: The BBC Plan for Network Radio and Non-Metropolitan Broadcasting. BBC Publications, 1969.
[2] Broadcasting Policy; White Paper, 2 August 1966. C.(66) 125.

attributes of the now defunct pirate stations. As Hendy (2007, 14) states it was created by 'the exigencies of pop.'

The Light Programme was now renamed Radio 2, offering middle-of-the-road pop and rock music as well as folk, country, jazz, big-band music and light classics along with comedy and sport. Initially, Radio 1 and Radio 2 simulcast programming during many periods of the day. From 1970 however, the stations began to develop separate identities for different audiences, with Radio 1 targeting the youth culture audience in popular music while Radio 2 tried to keep step with the musical tastes of people in their 40s and 50s. Roger Scott, who was Controller of Radio 1 and Radio 2 when both stations launched, admitted to being unsure at that time as to what actually constituted 'pop music' in order to differentiate between the two stations. It was not defined in the government White Paper of 1966[3] but was left for the BBC to define. Scott's solution was to decide it would be different to what he himself described as 'sweet music,' an umbrella term for most of the music offered on the Light Programme (Scott 1967). Given the opportunity to establish its own pop music network, it would appear that the BBC was still unsure about the formula that had made the pirates so successful, even stumbling over definitions.

The Third Programme, established as the main cultural station offering classical music, drama and talks and experimental programmes, was named Radio 3. To be more precise 'the *Third Network*, the umbrella title for the Music Programme, the *Third* and the other various sport and education programmes that shared the frequencies with them, became Radio 3' (Carpenter 1996, 247), although the new channel would have a narrower remit. In fact, the wrangling over exactly what would be broadcast on Radio 3 and what sections of its content might shift to Radio 4 continued up until 4 April 1970 when the new Radio 3 network plans came into effect, namely, a 'mainly classical music repertoire with some contemporary music and drama' (Carpenter ibid, 257).

The Home Service had been the home of spoken word programmes including news and current affairs, discussion, drama and features. On becoming Radio 4, it was to retain its familiar range of programming and indeed for a while found it difficult to shake off its old name which survived for a year or more in the form of the on-air announcement: 'This is Radio 4, the Home Service.' In its transition from the Home Service, Radio 4 was the network that underwent the least change in its output,

---

[3] ibid.

many of the old programmes remained the same as Hendy (2007, 15) notes:

> The sheer range and mix of programmes was impressive. But it was a
> menu barely distinguishable from that published a week before under
> the Home Service moniker—barely distinguishable, indeed, from anything
> printed in the *Radio Times* a decade or more before.

By 1970 however, Radio 4 had begun to shape its own identity, transforming from its old Home Service style of mixed speech and music to a 'wholly speech oriented station' (Elmes 2007, 16) where journalism would be allowed to bloom and current affairs would have a more important, central role alongside the existing drama, comedy, science and arts output coverage, so, for example, programmes such as *The World Tonight*, *PM* and *Start The Week* all began in this year.

Briggs (1995, 577) claims that apart from Radio 1, which really was completely new, 'bigger changes had already taken place in BBC radio in the years between 1964 and 1967 than took place in 1967 itself.' This may be true in the sense that the Light Programme had begun its battle with the pirates by slowly introducing some popular music programming—leading some to describe it as a 'youth club of the airwaves' (Chapman 1992, 1)—and with the introduction of the Music Programme as part of the Network Three strand on the Third Programme, but it failed to recognise that the changes resulting from the restructuring of 1967 had far greater implications for the future of radio in the BBC. There is no doubt the reorganisation of BBC radio services in 1967 was driven primarily by the hunger for pop music amongst the next generation of the BBC's potential audience. It could be argued that this drive was spurred on by a fear on two fronts: the fear of losing that audience and the fear of that audience being captured by any potential competitor. The pirates had been removed from the broadcasting arena, but they could always reappear or competition in another format could also emerge. The changes of 1967 illustrated a new direction for radio and a new strategy process largely inspired by the threat of competition:

> The attention being given to Radio 1 at the end of September 1967
> spoke of a new dawn in British broadcasting. The role of radio within
> the BBC's range of services seemed to be changing in fundamental ways,
> and the corporation's historic commitment to speech radio suddenly looked
> vulnerable. (Hendy 2007, 14)

The rearrangement of the BBC's national services did not just mark a more important position for popular music, it also revealed a new dimension to programming. The previous model of networks delivering a wide range of content was now abandoned in favour of more niche, audience-focused content, as the Managing Director of BBC Radio, Ian Trethowan (1970, 6), noted:

> The basic change in radio is that mixed programming no longer applies, listeners seek the convenience of predictable networks.

As well as a more prominent role for pop music and a greater desire to appeal to the youth section of the audience, the changes to national radio demonstrated a new commitment to *narrowcasting*—namely the targeting of specific content to specific audience sectors, a trend which prevails to the present day for most radio broadcasters.

BBC Local Radio

An interesting notion to rear its head in the 1960s was that of local radio, and it was one which was to assume a central role in the broadcasting policies of the latter half of the decade. Notions of local radio broadcasting suggested it implied a 'localised provision of news, music and crucially, democracy' (Wright 1982), so it was not based on the simple music formats of the pirates, although its perceived role as a bulwark to the threat of the pirates was certainly also an important consideration. The idea of local radio was initially considered at governmental level in a White Paper in 1962[4] and subsequently in a second White Paper later in the same year.[5] In December 1966 the Labour government announced its decision to go ahead with local radio in another White Paper, the same White Paper that authorised the BBC to begin a new pop music network.[6] However, as far as local broadcasting was concerned it was deemed only a partial victory, with local radio to be launched on an experimental basis and run as a joint operation by the BBC and local authorities.[7] The experiment began on 8 November 1967 with the opening of BBC Radio Leicester with BBC

---

[4] Broadcasting Policy; White Paper, July 1962. Cmd 1770.
[5] Broadcasting Policy; White Paper, 13 November 1962. C.(62) 181.
[6] Broadcasting Policy; White Paper, 20 December 1966. Cmd 3169.
[7] White Paper Christmas, *The Spectator*, 23 December 1966.

Radios Sheffield, Merseyside, Nottingham, Brighton, Stoke on Trent, Leeds and Durham following next in line.

The development of local radio in Britain was expected to be more than simply a copy of what the pirates had been doing or of what was apparent in the successful American model as extolled by Frank Gillard[8] following a trip to the USA in 1954 where he celebrated its ability to speak to listeners as a 'familiar friend and neighbour.'[9] It was thought local radio must have significant and positive effects on local political communication, and one of the key driving forces at governmental level was the need to promote greater public interest and participation in the system of local government. The White Paper of 1966[10] had identified this as a central part of the raison d'etre of the proposed new system of local sound broadcasting which was charged with 'fostering a greater public awareness of local affairs and involvement in the community' (Wright 1982). In fact, local broadcasting was seen as a crucial component to social integration theories of the period (Powell 1965).

As well as considering the future of its national services, the BBC was at the forefront of the emerging local radio vanguard; in fact Briggs (1995, 619) argues the main thrust for it came from inside the BBC. He mentions Gillard as a leading advocate of local radio within the BBC and cites the latter's document from 1955—'An Extension of Regional Broadcasting'[11]—as the first BBC internal document proposing regional expansion but stopping just short of vaunting a local radio system largely on the grounds of cost. The BBC did put a proposal to the Pilkington Committee in February 1961 for around 80 local stations which they could deliver but with the warning that any competition from commercial stations in this field would lead to problems of increased cost for the corporation. The main envisaged danger was that commercial local broadcasting would be networked and blanket commercial sound broadcasting would come in sneakily through the back door.[12]

The commercial sector was quick to jump on the local radio bandwagon too. By autumn 1960, over one hundred commercial radio companies, mostly backed by local press, had been formed in anticipation of a

---

[8] BBC Director of Radio 1964–1969.

[9] Radio in the USA: A Visitor's View, Frank Gillard, 6 July 1954. BBC WAC: E15/75.

[10] Broadcasting Policy; White Paper, 20 December 1966. Cmd 3169.

[11] An Extension of Regional Broadcasting, 28 February 1955. BBC WAC: R34/731/5.

[12] Report of the Committee on Broadcasting: Memoranda Submitted to the Committee, Vol 1: The Future of Sound Broadcasting: Local Broadcasting. Cmnd 1819 (1962).

new committee on broadcasting which might 'favour the local dimension' (Briggs ibid, 629). The BBC was quite fearful of this happening and had identified its possible impact, believing it would be lucky to retain just over one third of the total available audience for radio if local commercial companies were allowed to go ahead.[13] Commercial pressure continued with larger forces such as Pye—the radio manufacturing company—joining the campaign.[14] The BBC's reaction was to rely on its old method of castigating the commercial companies by addressing the type of programming they would offer and how the BBC's would be more virtuous. Speaking in December 1963, Gillard reflected:

> BBC stations would be more than mere juke-boxes but instead would have programmes reflecting current affairs and local interests, issues and cares.[15]

A BBC statement claimed the aim of its local stations would be to help build a vigorous and satisfying local life with a wide progressive outlook, essentially suggesting the BBC would be best at serving local communities:

> The BBC believes that what broadcasting has done for the national community over the years, it could also do for the local community... BBC local stations would devote themselves to local issues and interests, to provide a service which would effectively enlarge the range of broadcasting in Britain and meet a genuine need in each modern community. Everything of real concern in community life would be reflected and covered in the programmes.[16]

Indeed it was no surprise to the BBC when the Postmaster General John Stonehouse wrote to BBC Chairman Lord Hill in August 1969 telling him the government now authorised 'the provision by the BBC of a general and permanent service of local radio broadly on the lines proposed by the corporation.'[17] The Postmaster General's decision resulted in the

---

[13] BBC Marriot Committee Report on Area and Local Broadcasting, 1 September 1959. BBC WAC: R34/1585/1, Local Radio.
[14] A Plan for Local Broadcasting in Britain. Pye Telecommunications Ltd, October 1960.
[15] Interview with Frank Gillard. *Yorkshire Post*, 11 December 1963.
[16] Local Radio in the Public Interest: The BBC's Plan, February 1966, BBC Publications.
[17] Stonehouse to Hill, 13 August 1969. BBC WAC: R78/610/1.

BBC being given permission to increase the number of stations from the existing eight to twenty.

Linfoot (2011, 185) argues that the future for BBC Local Radio by the end of the 1960s was increasingly becoming tied to party politics. The Labour government (1964–1970) was happy to secure the future of BBC local radio and increase the number of stations, but with the coming to power of a Conservative government in 1970, the arrival of commercial radio was inevitable. A new White Paper in March 1971 laid out plans to create up to 60 new local commercial stations across the UK,[18] the BBC's monopoly in local broadcasting appeared to be under threat.

Broadcasting in the Seventies

It is important at this point to consider the BBC policy document, *Broadcasting in the Seventies*,[19] which was published in 1969 and is to this day seen as a seminal document in the history of BBC Radio. Jenny Abramsky[20] (2002) describes it as one of the most controversial documents ever produced by BBC Radio and it generated considerable controversy largely because it developed further what Chignell (2011, 92) refers to as 'generic' broadcasting or 'format' broadcasting, namely a model which targets specific audience groups. This was a model which was pioneered by commercial stations and which the BBC felt the need to embrace and one which would go on to dominate UK radio up to the present day. In effect, *Broadcasting in the Seventies* may be rightly seen as a landmark document but it should also be remembered that its driving imperative emanated from a broadcasting model constructed by the BBC's competitors.

The document begins by highlighting how the BBC has moved into new territories, including Radio 1 and local radio, but reveals that this has exposed a weakness:

> These changes, however, have been grafted piecemeal on to a tree planted in an earlier age of broadcasting, and we have now looked at the radio services as a whole, to see how they might be rationalised and reshaped to serve the audiences of the seventies.[21]

[18] White Paper: An Alternative Service of Radio Broadcasting, 1971 (Cmd 4636).

[19] *Broadcasting in the Seventies: The BBC Plan for Network Radio and Non-Metropolitan Broadcasting*, BBC Publications, 1969.

[20] BBC Director of Radio 1999–2006, Director of Audio and Music 2006–2008.

[21] ibid., p.2.

The document goes on to describe how the traditional broadcasting model based on the principle of mixed programming on a single channel is anachronistic and listeners now expect radio to be based on a different principle:

> ... that of the specialised network, offering a continuous stream of one particular type of programme and meeting one particular interest.[22]

A priority would therefore be a clearer separation between Radio 1 and Radio 2. There was also a proposal to realign Radio 3 and Radio 4. On Radio 3, the continued separate labels of Music Programme and Third Programme would disappear with the entire output of the network put under the single heading of Radio 3, concentrating wholly on music and the arts. The aim was also to confirm Radio 4 as a speech network, the home of factual programmes, documentaries, current affairs as well as news.[23]

As for local radio, after embracing the local radio experiment of 1967 there was now a belief that local radio was not only viable, but an integral part of any broadcasting system. This was referred to as 'Radio 5'—a system of some 40 local stations, broadcasting local news and information and a whole range of community programmes.[24] The document speaks of the need to develop local radio as 'a vital part of the BBC's services,' but interestingly states this is not done in an attempt to stake a claim as a sole provider of such a service by insisting 'we make no claim for monopoly,'[25] aware that competition may be on the horizon at local level. This threat of local competition is nevertheless not considered to be in any way critical as the document confidently predicts that as the BBC has matched up to competition in television 'it could do so in radio.'[26]

Jenny Abramsky had only recently joined the organisation at this time and noticed 'a narrowing of BBC Radio's ambition' (Abramsky 2002) as revealed in the document. Although in many ways visionary by creating a shape for public service radio that rendered it more audience-focused and ultimately more competitive, Abramsky also concludes that the all-pervasive message from the document was that 'radio had to change since

[22] ibid., p.4.
[23] ibid., p.4.
[24] ibid., p.13.
[25] ibid., p.6.
[26] ibid., p.6.

we were now in a television age, the core belief was that radio still had a role in some areas but for most people radio was merely a daytime medium' (Abramsky ibid). Abramsky also claims that around this time there was a feeling that 'radio was no longer central to the future of the BBC' and indeed some insiders were even predicting its 'ultimate demise by the end of the century' (Bridson 1971, 333).

In many ways *Broadcasting in the Seventies* represented a landmark document for the BBC and Briggs offers a comprehensive exegesis, devoting an entire chapter to it (Briggs 1995, 721). The key changes in the content of the BBC's national radio stations and the development of BBC local radio came about as a result of implementing the policy document, resulting in a greater coherency of programming. In a wider, more theoretical context it is important in shifting the BBC's approach away from mixed programming towards more niche radio broadcasting. The creation of the four networks in 1967 certainly initiated this process, but it was *Broadcasting in the Seventies* which consolidated it by ensuring each network now had a unique and separate identity. In many ways one can perceive this document as the cradle of narrowcasting[27] within the BBC, namely, 'the customized version of broadcasting that targets information to a specific, narrowly defined group of recipients' (Jones 2003, 337). It is also interesting to note with the benefit of hindsight how Abramsky's fears proved unfounded and that radio was to undergo a renaissance a little over 20 years later although it is easy to see how at this point in history her argument was persuasive.

Certainly the 1970s were perceived within the BBC to be the decade of massive anticipated change. Speaking as the new dawn approached, the BBC Chairman, Lord Hill, was able to herald the new era for radio as one centred around further development of the specialised networks like Radio 1 and Radio 3 but also a prominent position for local radio:

> Non metropolitan broadcasting forms an essential part of our service.
> The biggest development in radio over the next decade will be localism,
> we hope every major city and community will have its own radio station.
> The local radio experiment has been a great success (Hill 1969).

Broadcasting in the Seventies is an important document because it is the first major policy document which sets down how BBC Radio must adapt its radio services in order to deal with competition and accepts that the

---

[27] Perhaps a better term for 'generic' broadcasting or 'format' broadcasting (Chignell 2011) by virtue of the fact that the word 'broadcasting' itself is omitted.

BBC's position may indeed succumb to a challenge from other sources. It highlights how the BBC has recognised the threat of past competition—and, more importantly, of possible future competition—but also illustrates how, by embracing the broadcasting models exemplified by competitors, the BBC has been forced to move from a Reithian notion of 'mixed programming' to one of specialised, audience-focused services, if it was to compete successfully, and as Chignell (2011, 95) states, the document 'left nothing to the imagination and no room at all for obsfucation,' which emphasised how significant a policy change it represented. Finally, the BBC had identified its weaknesses vis-à-vis its commercial competitors and highlighted its response in a policy document. Localism and specialisation would become the major forces in the 1970s for the BBC. They would also create the level playing field upon which commercial agitators would finally get their chance to once again take on the BBC, not just at a local level but eventually also at a national level.

The Competition: A New Radio Entity

By the end of the 1960s, with the pirate stations no longer visible or audible on the radio horizon, the BBC faced virtually no actual direct radio competition from competitors with the exception of Radio Luxembourg which still managed to maintain a decent audience share against BBC Radio by the mid-1970s. Any notion of monopoly within the BBC after the launch of its new national and local services was however to be short lived. Briggs (1995, 638) suggests that by the late 1960s the BBC was beginning to 'reconcile itself to eventual competition' and that its radio monopoly was not secure. In reality, it was a change in the political landscape which brought the threat home. In its manifesto for the 1970 General Election, the Conservative Party spelt out its unambiguous plans for radio in the UK:

> We believe that people are as entitled to an alternative radio service
> as to an alternative television service. We will permit local private
> enterprise radio under the general supervision of an independent
> broadcasting authority. Local institutions, particularly local newspapers,
> will have the opportunity of a stake in local radio, which we want to see
> closely associated with the local community.[28]

---

[28] 1970 Conservative Party General Election Manifesto.

On winning the election, Prime Minister Edward Heath[29] set these plans in motion and the resulting piece of legislation was the Sound Broadcasting Act 1972 which established for the first time a legal position for competitors to the BBC. At the same time, the Independent Television Authority (ITA) accordingly changed its name to the Independent Broadcasting Authority (IBA) and would also act as regulator for the new tranche of sound broadcasting. The Act received Royal Assent on 12 June 1972 and on 19 June Minister for Posts and Communications, Christopher Chataway, announced the location of the first wave of 26 Independent Local Radio (ILR) stations.

Independent Local Radio (ILR)

It was clear from the 1970 Conservative Party election manifesto and the subsequent White Paper entitled An Alternative Service of Broadcasting[30] that the new proposed system of local commercial radio was not going to be based on a free market model but rather one which espoused public service attributes, catering for local news and information needs and thus not at all similar to the models of commercial broadcasting which had come and gone since the 1930s. Amongst the proponents of commercial radio in the UK, many shared this ideal of how local commercial radio should operate. The Local Radio Association, which had been campaigning for many years, had refined its own position in a document it published in 1970 called The Shape of Local Radio[31] although not everyone shared this new analysis. As Stoller (2010, 29) points out, another promoter of commercial radio, Hughie Green, was 'vocal in his opposition.' So we begin to see early divisions set in amongst commercial radio champions who, on learning their hopes may become reality, begin to fragment over what shape it should take.

The White Paper proposed the continuation of the BBC largely unchanged with the provision by the BBC of 20 local radio stations, funded by the licence fee. The IBA was to provide advertising-supported local radio from up to 60 stations, and any uncertainties regarding their relationship with the BBC were laid to rest with an explicit reference to

[29] In office 1970–1974.
[30] White Paper: An Alternative Service of Radio Broadcasting, 29 March 1971 (Cmnd 4636).
[31] The Economist, 10 October 1970.

the newly named ILR stations competing directly with Radio 1 and Radio 2 rather than with BBC local radio (Wray 2009), in effect laying down the model of competitive broadcasting which would permeate the radio industry for the next two decades. The passing of the subsequent Sound Broadcasting Act (1972) finally gave full legal status to an opposing model of radio broadcasting to that of the BBC, and just over a year later, in October 1973, this model made the transition from the statute book to the studio with the launch of LBC in London.[32] LBC had won the news franchise for London and was followed a week later by Capital Radio[33] which won the London general and entertainment franchise.[34]

LBC's objective was to provide a news station for London as well as supply a news service for the new commercial stations, known as Independent Radio News (IRN). LBC did not start well, 'failing to attract anything like enough advertising' (Stoller 2010, 55) which was the primary source of income for the new stations. In these early years, LBC hastily went through a number of changes to its programming and staffing in an attempt to stem losses and find the programming and scheduling formula that would increase audience figures and therefore boost advertising revenue. On a number of occasions Chief Executive Bill Hutton envisaged the only way out of financial disaster was to rely on regular, complete overhauls of resources and output—what were in effect rescue plans.[35]

Capital Radio perhaps had an easier start simply because of its content, that is, a daytime mix of popular music and speech, something that was to characterise all other ILR stations, and this is why Stoller (ibid) suggests that Capital's start date, 16 October 1973, was arguably 'the real start of ILR in the UK' as this was the model that would prevail for future stations. Stoller does go on to say that despite Capital's relative success compared to LBC, it would be Radio Clyde in Glasgow which would set the standard for success when it began broadcasting on 31 December 1973, offering 'high grade, popular radio, mixing a skilfully chosen selection of music with ambitious news and features programming' (Stoller ibid, 60). Certainly for the two London stations spearheading commercial radio in the UK it was not an easy launch. In addition to the challenge of securing

[32] The London Broadcasting Company (LBC) began broadcasting on 8 October 1973.
[33] Capital Radio began broadcasting on 16 October 1973.
[34] The first five franchises were for London News (LBC 8 October 1973), London General (Capital Radio 16 October 1973), Glasgow (Radio Clyde 31 December 1973), Birmingham (BMRB 19 February 1974) and Manchester (Piccadilly Radio 2 April 1974).
[35] Rescue plan for LBC to be put to staff. *Financial Times*, 23 January 1974

advertisers, both stations received poor press reaction. While Radio 2 may have been seen as Capital's natural competitor and Radio 4 LBC's, competitive tensions between LBC and Capital themselves were a feature of the weeks and months following launch. In London, the presence of two new radio stations meant they were competing with each other for audience and revenue and found themselves both party to common disputes over the coverage of various events in the capital.[36] For the new ILR stations the competitor was not just BBC local and national radio but other ILR stations. In order to convert this division into cooperation, the first five companies met in March 1973 and formed the Association of Independent Radio Contractors (AIRC) to handle issues where 'joint work was desirable,' thus marking an attempt to foster better relations among ILR stations, particularly in their joint struggle against the BBC.

Stoller (ibid, 68) claims 1974–1976 were the pioneer years of independent radio. Stations had begun bonding with larger audiences and by early 1975 there were nine stations broadcasting.[37] An air of arrogant confidence became detectable among the ILR stations. The Managing Director of BRMB, David Pinnell, speaking in 1974 insisted:

> We are putting up a good fight against strong competition particularly
> from Radio 1 and Radio Luxembourg, but as far as community
> broadcasting is concerned we are way ahead of BBC local radio here.[38]

Even by 1976 however, the ILR stations still felt they were facing an uphill struggle. With the London stations still 'floundering against the BBC' (Hendy 2007, 139), Capital rethought its music policy, offering more pop music and less talk, for which it was criticised by the IBA, but Capital claimed this was necessary in order to boost audience appeal. It was a move that paid dividends so that by the middle of 1977 Capital had caught up with Radio 1, with audience figures showing only a fraction of a point separating the two after Capital had increased its audience by 100,000 in just six months.[39] Only one year later, Capital had finally

---

[36] Stoller (2010, 56) cites the tensions over the wedding of Princess Anne and Mark Phillips on 14 November 1973 which required intervention by the IBA.

[37] LBC, Capital Radio (both London), Radio Clyde (Glasgow), BRMB (Birmingham), Piccadilly Radio (Manchester), Metro Radio (Newcastle), Swansea Sound (Swansea), Radio Hallam (Sheffield), Radio City (Liverpool) and Radio Forth (Edinburgh).

[38] *Ad Week*, 12 July 1974.

[39] *Evening Standard*, 7 July 1977.

managed to topple the BBC by taking over the mantle of the most popular radio station in London.[40] Naturally the BBC fought back by refocusing Radio 1 even more on the music charts which did affect Capital in the short term, but the London station in turn reinvented itself by eschewing teenage music trends such as punk and honing in more on chart music,[41] the staple of those who were going to buy the products which companies were willing to pay to promote on air and the results were apparent:

> Commercial radio is now enjoying its first real boom in both audiences and revenue since it started two and a half years ago... Many of the ILRs are the most popular stations in their area, beating Radio 1 and Radio 2. The turning point for commercial radio was Autumn 1975 when advertising picked up.[42]

Indeed by 1982 Capital had recovered its audience figures and also become recognised as Britain's richest radio station.[43]

The 1970s closed on a good commercial note for ILR stations with many now trading profitably. A cloud on the horizon came in the form of the Annan Report 1977[44] which although mostly concerning television, introduced potential conflicts from that medium in the shape of Channel 4 and Breakfast TV. We also witness at this time an emergent degree of cooperation between the BBC and the ILR stations. The Home Office set up a Local Radio Working Party (LRWP) in wake of the Callaghan government's White Paper of 1978,[45] and it was at this point where the IBA and BBC Radio engaged with each other, albeit on both competitive and cooperative levels. The LRWP discussed matters such as future frequency allocations, particularly the future of local radio for both the BBC and IBA, and produced a series of three reports which Linfoot (2011) argues 'effectively set the pattern for the immediate growth of local radio.' Linfoot (ibid) cites an interview with Michael Barton[46] where the latter

---

[40] *Evening Standard*, 29 June 1978.

[41] It is important to remember that at this time Capital was also offering drama and classical music content which was competing successfully with Radio 3 and Radio 4 (Seaton 2015, 97).

[42] Commercial Radio: the 19 stations that showed why the cynics were wrong. *Campaign magazine*, 2 July 1976.

[43] *Financial Weekly*, 13 August 1982.

[44] Annan Report on the Future of Broadcasting, 24 February 1977 (Cmnd 6753).

[45] White Paper on the Future of Broadcasting, July 1978 (PREM 16/1525).

[46] Controller BBC Local Radio 1976–1988.

claims the workings of the LRWP 'witnessed a period of much closer relations between the BBC and the IBA' and it represents the first real 'forum and means by which the BBC and the IBA could agree frequency allocation for the next wave of local stations' (Linfoot ibid). Once this issue was settled, the LRWP became redundant and a period of close relations came to an end. The degree of cooperation exposed by the workings of the LRWP may represent a subtle change of direction, particularly for the BBC, but it should be tempered by Barton's observation only a few years before when he was able to find comfort in the knowledge that:

> Commercial radio has given us an added stimulus to the job we are doing; but we believe... that we have different roles to play (Barton 1976).

The challenge for the ILR stations had always been to position themselves between the opposing forces of public service broadcasting and market forces, all while under the constant regulatory gaze of the IBA. Steps taken by Capital and other stations to reconcile local output and distinctive content with the desperate need to promote pop music programming in order to attract both audience and advertisers, were proving successful. This was the point at which it is possible to identify the start of the move across the Rubicon 'as independent radio began to aspire once more to be commercial radio' (Stoller 2010, 123).

'Independent' v 'Commercial'

The distinction between 'independent' radio and 'commercial' radio requires analysis at this point as it is important to understand how such a peculiar hybrid model emerged. Commercially funded radio, established in 1973 and regulated by the IBA, is referred to both in formal documentation and academic appraisals as 'Independent Local Radio' (ILR), certainly up until 1990 when the term 'commercial radio' became the norm. IBA regulations bound franchise bidders to a strong public service remit, encouraging 'programmes of merit'[47] including education, religion, meaningful speech and a range of music. This was combined with draconian curtailment of their commercial activities and strict monitoring,[48] all of which bore little resemblance to the previous forms of commercial activity of the

---

[47] Notes on Independent Local Radio. IBA, 12 July 1972.
[48] ibid.

1930s or 1960s. The BBC had by this stage long shaken off its own traditional Reithian restraints and no evidence of 'Reithianism'—by which I mean a strict adherence to prescriptive regulation based on safeguarding perceived standards—had been evident in the activities of the BBC's competitors up until this point. So, it was an interesting and surprising philosophy to regain currency in the new independent radio model. Street (2002, 118) believes it was not lost on those who tuned in how conservative the output of the new ILR stations was:

> It was to be one of the ironies of the arrival of land based commercial radio in Britain that over-regulation diluted the character of this new 'voice', far from free-style mavericks who had fought so hard in the 1930s and 1960s for this moment, the first era of ILR was a curiously Reithian affair.

There now appeared to be an entirely different persona to what had been the accepted model of the commercial sector, but why did this come about? (Street 2006, 208) suggests the answer may be found in the 'conveniently vague' nature of the service as outlined in the Sound Broadcasting Act 1972:

> The two stools between which independent radio therefore fell were on one hand the suspicion that regulation would be interpreted so lightly as to be non-existent and on the other that over caution on the part of the new regulator would stifle the medium's financial development and prevent it from being truly independent.

The Conservative government which came to power at the 1970 general election, confounding the opinion polls, may not have fully considered the intricacies of its pledge to introduce commercial radio and instead delivered something which eschewed full-blown commercialism and was cast in a detailed set of statutory regulations. Stoller (2010, 2) says this model, which fused notions of private radio funded by advertising but delivering public service output, was peculiar to the UK, but this ignores a similar model for radio and television employed by the Irish national broadcaster, Raidió Teilifís Éireann (RTÉ), which applied a similar framework to its radio and television strategies (Corcoran 2004). Wall (2000) suggests this notion of 'independence' as given in the government White Paper on broadcasting of 1971 was not necessarily used as a foil to commercialism but was instead designed to allow the new sector to be presented as:

> *independent* from the BBC's public corporate status, *independent* from the BBC's programming monopoly and London-centric organisation, *independent* from the commercial pressures of media conglomerates like those in North America and *independent* from the influence of US-originated popular culture.

Certainly the eventual form of commercial radio which was introduced was largely unfamiliar to the accepted model of commercial radio with its core characteristic of 'homogenous programming centred around pop music' (Shelley and Winck 1995, 117). 'Independent' was evidently the key word as opposed to 'commercial' and it was this distinction, together with the responsibility of stations to answer to audiences rather than share-holders, that led to the crises of survival in the first few years of the ILR stations' existence. Even the BBC had to be careful when using the terms 'independent' and 'commercial' interchangeably as Director General Ian Trethowan informed a meeting of the Board of Governors that IBA Chairman, Lady Plowden, had remonstrated with the BBC regarding the word 'commercial,' claiming that it offended the IBA to be described in terms that implied they were not engaged with public service broadcast-ing.[49] It is also important to remember that the new stations founded from the Sound Broadcasting Act 1972 were clearly 'characterized by a public service broadcasting ethos' (McCain and Lowe 1990) and until the late 1980s ILR stations had to 'make public service programmes in order to secure their licences to broadcast in a regulated system' (Seaton 2015, 96).

As a point of semantics, it is easy to define the difference in the terms 'independent' radio and 'commercial' radio as they represent two quite distinct models of radio broadcasting. In terms of popular culture, both terms were used interchangeably to describe that model of sound broad-casting emanating from a body which was not the BBC. At an academic level however, it is important to adhere to a distinction between the two, which then begs the question: where is the precise point when one becomes the other? Certainly the 'independent' model prevailed in the

---

[49] Board of Governors meeting minutes, 7 September 1978. BBC WAC: Commercial Radio General R92/301/1.

1970s, but the emergence of the 'commercial' model can be traced back to the 1984 Heathrow Conference.

Heathrow Conference 1984

The expansion in the number of ILR stations following the Broadcasting Act 1982 meant that by the tenth anniversary of independent radio in the UK in October 1983, there were 38 stations on air across the country, but the rules under which they operated remained exactly the same despite the fact that the core policy of the Thatcher government seemed to promote commercial freedom and 'free market' economics. This did not seem to apply to the UK radio sector and those at the helm thought this was precisely the sort of environment which could make them ultimately more profitable. As Wray (2010) notes:

> Frustration and concern over the restricted business model was mounting; the desire to create radio for segmented audiences to increase revenue and ratings was becoming more urgent… yet at this point the sound of Independent Local Radio had changed little.

It was the chairmen of the original 19 companies who took the first initiative early in 1984, forming themselves into a lobbying group in order to liberate the independent radio sector from the shackles of intervention which it believed was impeding its commercial development potential. It was agreed by the chairmen and by the AIRC council to convene a special meeting to lay out their grievances and demands which they would then take to both the IBA and the government. That meeting took place on 23 June 1984 at the Sheraton Skyline Hotel at Heathrow airport and became known as the Heathrow Conference. Stoller (2010, 153) claims the impact of the Heathrow Conference was threefold:

> It focussed the strain and worries of the ILR companies on a challenge to the fundamental conception of ILR as *independent* rather than *commercial* radio… it brought ILR as an industry into direct contact with the government for the first time… it changed the relationship between the companies and their regulator.

Street (2002, 125) describes the conference as a 'Council of War' and this is an apt description. The ILR chairmen were unhappy with the restrictions placed upon them and were going to challenge not only the IBA but also the government. From the conference came four key resolutions which formed their 'Council of War' strategy:

1. Make public the industry's frustration at IBA over-regulation.
2. Demand an early, substantial cut in rentals.
3. Press the government for new legislation on commercial radio.
4. Commission an independent report on the potential for creating more stations.[50]

The conference was a success as the IBA did listen and within only a few months many of the IBA rules had been dispensed with. A good example of this was the rule regarding commercial sponsorship. From 1985, the ILR Network Chart Show[51] was permitted to attract more outside commercial sponsorship of the programme. This was a massive change in IBA policy, especially considering the initial proposal for a network chart show to compete with the BBC's had already been refused only as recently as 1979. The Network Chart Show could now go ahead against the BBC and attract hugely valuable sponsorship and advertising, thus becoming a paragon of 'commercial' rather than 'independent' radio.

The achievements of the Heathrow Conference would have long-term consequences for the independent sector and were even crucial to the survival of the industry. Stoller (2010, 4) does not underestimate its significance when he says this was the point where the independent radio industry 'broke with its regulated past and began to drive towards a commercial future' and Wray (2009) similarly concludes it changed 'the nature of commercial radio regulation and brought a new confidence to the people within it.' In the history of commercial radio in the UK, the Heathrow Conference marks the first step towards a future which would be truly commercial and fully enshrined by the 1990 Broadcasting Act. It also highlighted how cooperation among actors within commercial radio led to a much improved standing for the sector and arguably for the benefit of the entire radio industry, as well as heralding the end of competition with the BBC within the confines of a strict public service remit.

---

[50] These four key points are a summarised version of the six resolutions passed at Heathrow; see Appendix.

[51] The Network Chart Show was a Top 30 countdown show launched across the ILR network on 30 September 1984, competing directly with the Radio 1 Top 40 countdown.

THE BBC REACTION: EMPHASIS ON LOCAL

For BBC Radio the main concerns which prevailed throughout the 1970s and 1980s centred around what might happen to its own radio portfolio and how the ILR stations might impact on it. Even after its launch and early success, the future of Radio 1 remained uncertain and there were even fears that it was this station, by its very nature, which might be snatched from the BBC and moved to the commercial sector. But in January 1971 this fear was alleviated when the Minister for Posts and Telecommunications, Christopher Chataway, announced that Radio 1 and the local BBC stations would remain firmly within the BBC and that ILR would have an entirely separate existence.[52]

As for the arrival of a new and eager element of competition in the UK radio industry, the BBC did not seem unduly perturbed. The *BBC Handbook 1975*[53] laments the ending of the BBC's monopoly in radio with the arrival of the first ILR stations but concludes that BBC Radio's own audience figures seemed to be unaffected by this and boasts that at the end of the year under review more people were listening to BBC Radio than at the end of the previous year. The *BBC Handbook* for the following year, 1976,[54] admits a marked increase in public interest for ILR but also claims that despite this, the audience for the BBC local stations rose sharply, citing a statistic which showed that in last quarter of 1974 over 1,800,000 people on average listened to BBC local radio each day, an increase of 50% on the previous year. The Controller of Radio 1 and Radio 2, Douglas Muggeridge, felt able to report on the approach of the first anniversary of commercial radio that the BBC had experienced:

> ... no loss of audience, if anything our figures are running slightly higher than last year. The large audiences quoted by the commercial stations bear no relation to our figures.[55]

Of course the commercial stations completely disagreed with Muggeridge and claimed his figures were not based on accurate research methods.[56]

---

[52] *The Listener*, 11 February 1971.
[53] Incorporating the Annual Report and Accounts for 1973–74.
[54] Incorporating the Annual Report and Accounts for 1974–75.
[55] BBC Press Release, 24 Sept 1974.
[56] Audience Research, Commercial Radio. BBC WAC: R78/3, 861/1.

Unsurprisingly, the BBC and the commercial sector set off on a path of overt competition without any desire on either side to seek rapprochement, never mind any minor degree of cooperation during the first few years of their dual existence. As it had done with the IBC back in the 1930s, the BBC was constantly watching the plans and activities of the new commercial stations by asking staff to keep their ears to the ground.[57] It also became worried when it thought ILR stations were beginning to act in concert in order to gain publicity as a group in an attempt to compete directly with the BBC:

> Papers report several commercial radio stations would be broadcasting the last book of the Pallisers[58] on the night BBC TV should have been showing it. LBC took a half page advert in Evening News to announce it. This is a big publicity coup for commercial stations.[59]

Any publicity for the ILRs was bad enough, but when it came from inside the BBC itself then recriminations were not far behind as when a BBC Radio 4 programme *Midweek* included a discussion from the studio of the Edinburgh commercial station, Radio Forth, in May 1975. Alasdair Milne[60] thought:

> ...the promotion of Radio Forth odd to put it mildly. I have made enquiries... I do however regret that Radio Forth should have gained such free publicity from us.[61]

The commercial stations were not afraid to stand up to the BBC and certainly fought their corner, largely due to an unabated self-belief and a strong desire to compete with the latter. This was after all the culmination of decades of challenge to the superiority of what the BBC's rivals believed to be the former's monolithic monopoly. Simply because the BBC had been around for a very long time did not imply that only it had the

[57] Memo from David Lloyd James to Network Controllers, 28 May 1974. BBC WAC: Commercial Radio General, R92/301/1.

[58] *The Pallisers* was a 1974 BBC television adaptation of Anthony Trollope's Palliser novels.

[59] Memo from Chief Publicity Officer, Radio (Michael Colley) to DPR, Recent Press Coverage, 15 July 1974. BBC WAC: Commercial Radio General, R92/301/1.

[60] A future Director General but as this point Director of Programmes, Television.

[61] Memo from Alasdair Milne (DPT) to Howard Newby (DPR). Radio Forth, 15 May 1975. BBC WAC: Commercial Radio General, R92/301/1.

wherewithal to provide exceptional programming and this was particularly felt at LBC where its editor proclaimed:

> The BBC has no divine right to expect an audience for its news simply because of its reputation for accuracy and balance built up over a large number of years.[62]

Chignell (2011, 139) in fact argues that the major innovations which came about in radio news broadcasting were made by commercial radio and highlight how LBC/IRN made 'a major contribution to radio journalism.'

As each new ILR station appeared on the map, it confidently announced its arrival with a remark on how it hoped to impact on the BBC. So when Metro Radio launched in Newcastle upon Tyne on 15 July 1974, it was able to state that it fully expected not just to take on BBC Radio Newcastle but actually win over the Radio 1 and Radio 4 audience in the North East.[63] Such bold confidence continued, and by 1976 the IBA heralded the completion of its first phase in the development of ILR by highlighting how it had established 19 stations, bringing a company to air every 6 weeks, reaching in excess of 25 million people each week on FM VHF and over 30 million on medium wave with around 12 million adults now listening to ILR each week.[64] Despite persistent financial difficulties the ILR stations felt they were snapping at the heels of the BBC both locally and nationally in terms of audience figures.

In the latter part of the 1970s we witness the first signs of a cordial communication between the BBC and the commercial sector. Previous to this, the relationship between the two had been quite confrontational and mistrusting with an absence of any element of reciprocity. In 1978 we see evidence of the first tentative steps towards some small degree of collaboration after the BBC Director of Radio, Aubrey Singer,[65] met with the Director General of the IBA, Sir Brian Young. Singer suggested to Young that there may be sufficient grounds for both to meet from time to time to discuss issues such as the siting of transmitters and allocation of wavelengths for each of their local stations. Young agreed, hoping that both

---

[62] Marshall Stewart, Editor LBC. *Campaign Magazine*, 22 February 1974.
[63] Interview with Metro Radio Managing Director Bruce Lewis. *Newcastle Journal*, 11 July 1974.
[64] IBA News Release, 9 April 1976.
[65] BBC Director of Radio, 1978–1982.

parties could identify some areas of common concern and perhaps from time to time opportunities might arise for cooperation. Cooperation was a new word to appear in the vocabulary of communication between the two parties and Singer highlighted it again when he replied: 'if you think there is any opportunity for sensible cooperation then I would be delighted to discuss it with you.'[66]

Despite a semblance of a spirit of cooperation, it was still only sentiment and had very little essence in practice, particularly within the BBC. A deep mistrust of the commercial sector remained even to the point of the BBC considering changing its bank when it learned that Barclays Bank was one of the biggest spenders on radio advertising.[67] Singer himself did not seem true to his words with Young, as he made clear in response to an approach from Capital Radio for BBC assistance in training independent radio staff:

> I don't think we can go with Capital Radio on this deal, simply because it allows them to get away with the impression that we are all one big industry and we are all exactly the same except they take advertising and we don't. We all know it is different to that.[68]

By 1980 a further attempt at dialogue was taking place between Singer and John Whitney, the Managing Director of Capital Radio. Both welcomed some form of dialogue on aspects of policy, with Whitney pointing out the only danger being 'an early meeting should find itself over concerned with matters on which we would not be likely to reach agreement.'[69] Singer proposed formal participation in the areas of joint research and the exchange of information on things like trade union relations and the issue of sports programming.[70] A formal lunch eventually took place between

[66] Correspondence between Aubrey Singer (Managing Director, BBC Radio) and Sir Brian Young (Director General, IBA), June/July 1978. BBC WAC: Commercial Radio General, R92/301/1.

[67] Memo from Controller Radio 2, Charles McLelland, to Managing Director Radio, 5 December 1978. BBC WAC: Commercial Radio General, R92/301/1.

[68] Memo from Aubrey Singer (Managing Director, BBC Radio) to Noble Wilson (Controller International Relations) Training Independent Radio Staff, 22 June 1979. BBC WAC: Commercial Radio General, R92/301/1.

[69] Correspondence between Aubrey Singer (Managing Director, BBC Radio) and John Whitney (Managing Director, Capital Radio) February/March 1980. BBC WAC: Commercial Radio General, R92/301/1.

[70] ibid.

leading figures in BBC Radio and leading figures from the commercial radio sector on 16 April 1980.[71] A number of key subjects were discussed, but for the purposes of identifying possible cooperation, the following are important to highlight as results achieved: the possibility of joint audience research, copyright (particularly the value of acting together against Phonographic Performance Limited (PPL) and Performing Rights Society (PRS)), archives (freedom of access to BBC archives in return for payment) and future liaison (the idea of a common radio group in the form of an informal meeting between the BBC and ILR managers).[72]

Whilst relations with the commercial sector remained formal and perhaps frosty, it was the BBC's relations with the government which were the most egregious. The Annan Committee reported on 24 February 1977[73] and one of its chief recommendations was the privatisation of the BBC's local radio stations. Concerned that the committee might have harsh suggestions for BBC local radio, the BBC underwent a huge analysis of its stations, surveying the possibilities open to them in the years to come and which it could use a central plank of the BBC's submission to the inquiry. The subsequent report became known as the Ennals Report,[74] and according to Linfoot (2011) it identified three distinct categories of proposals: a list of absolute priority for new stations, a list for station expansion in the present economic climate and a list of future developments. Ennals went on to suggest that existing stations had produced too much output of low quality, although he also concluded in his research that a 'sizeable proportion of people living in the present local radio areas feel involved with the station and a loyalty towards it.'[75] One of Ennals' most interesting conclusions was that he felt BBC local radio should not just appeal to minorities but aim 'to reach as many listeners as possible within a certain area,'[76] thus sowing the seeds of audience maximisation which had already been sown among some of those in the commercial

---

[71] Those present included BBC Managing Director Radio, BBC Deputy Managing Director Radio, BBC Chief Personnel Officer Radio, Managing Director Capital Radio, Managing Director Piccadilly Radio, Managing Director Radio City and Director Association of Independent Radio Contractors.

[72] Note on lunch between BBC and AIRC and ILR managers, 16 April 1980. BBC WAC: Commercial Radio General, R92/301/1.

[73] The Annan Report on the Future of Broadcasting, 24 February 1977.

[74] Named after report author Maurice Ennals (Station Manager at Radio Solent).

[75] Ennals Report on Local Radio Expansion 1975. BBC WAC: R102/22.

[76] ibid.

sector. Linfoot (ibid) argues that many of Ennals' proposals were eventually watered down by the BBC, but it would remain a vital template for the remainder of the decade in the next stages of local radio development.

Another report on local radio[77] was produced by the BBC governors in July 1978 and was devoted to challenging the Annan recommendations and promoting the value of local radio for the BBC. The report proposes the BBC's intentions for its local radio arm; these include an extension of coverage with the introduction of new stations and 24-hour broadcasting. It also goes on to highlight the fact that for the BBC, local broadcasting is entirely complementary to network radio and that the competitive position for the BBC can only be preserved by the combined strengths of Radio 1 and local radio, both of which were under threat. For the BBC, its very survival in the long term now seemed to be predicated on a dual aspect to its radio output, that is, a strong local service allied with a strong network presence. The report also considers the potential competitive effect of a national commercial network which was being advocated with some vigour by the IBA.

In March 1980 senior managers from BBC Radio met at Ditchley Park, a secluded country house in the heart of the Cotswolds, to look ahead to the next five to fifteen years in radio, that is, 1985 to 1995. The report which followed the conference[78] began by painting a picture of the changing radio environment with an emphasis on the suspected future competition from other media, particularly television in its various guises of local television, breakfast television, a fourth channel, cable television and teletext services[79] as well as video cassette recording. It also highlighted direct radio competition in the form of an expansion in commercial local radio. This would lead to a growing overlap and competition between stations in a particular area which it was thought, with a considerable degree of prescience, might encourage greater programming specialisation, a scenario in which the public would need to know exactly what choice of services the BBC was offering amongst 'all the other competing sounds which can be heard along the dial.'[80] The report proposes a substantive readjustment

[77] Report of the Local Radio Group of Governors, 20 July 1978. BBC WAC: Local Radio Development Governors Sub-Group, R92/67/1.

[78] Ditchley Conference Paper: Radio 1985–1995, March 1980. BBC WAC: Radio Services Policy, R92/2/1.

[79] For example, BBC Ceefax and ITV Oracle which had been running since 1974.

[80] Ditchley Conference Paper: Radio 1985–1995, March 1980. BBC WAC: Radio Services Policy, R92/2/1.

of the BBC's thinking on radio centred around a redefining of the very concept of public service radio in order to make it appropriate to a broadcasting world which is increasingly pluralist. In order to enable this, it was proposed that the existing number of national networks should be reduced while BBC local stations should receive greater support in the belief that local broadcasting was the primary growth area in radio. At this point we see the BBC focusing its energy on local radio. This was deemed the main area of competitive threat and therefore an expansion of the BBC's own local portfolio was necessary.[81]

Two years later BBC Radio produced its Radio Programme Strategy[82] for consideration by the Board of Management and Board of Governors which outlined its vision for the 1990s. Unlike the Ditchley Report, this report eschewed a restriction of the national networks but instead was keen on retaining a broad range of output and envisaged any contraction of existing services as potentially damaging to the BBC in competitive terms. The renewed long-term plan for BBC Radio was now based on, among other principles, maintaining a broad range of output, keeping the four national networks and a local radio network covering 95% of the UK and establishing clear channel identities.[83]

It would appear a series of policy shifts of emphasis had taken place over the course of this period, from national to local and back again to national in tandem with local. As the 1990s approached, another threat loomed on the horizon, namely, a national commercial station. The BBC was fairly certain that the commercial operators would now 'strive for their own national radio network.'[84]

CONCLUSION

From April 1971 the separate radio licence was abolished. The BBC licence fee would now cover television and radio which meant BBC Radio found itself in the dubious position of having to make claims on licence revenue, most of which would be directed to the more expensive and popular medium of television. On the face of it this could have sounded

---

[81] ibid.

[82] Radio Programme Strategy, September 1982. BBC WAC: Radio Services Policy R92/2/1.

[83] ibid.

[84] Note by Aubrey Singer (Managing Director, BBC Radio), 3 December 1980. BBC WAC Commercial Radio General R92/301/1.

the death knell for radio, but a number of factors contributed to its continued survival, namely; the pirates had shown that radio could compete with television at a certain level; local radio was seen as a saviour for the medium; commercial interests were keen on investing in local radio and *Broadcasting in the Seventies* spelt out the long-term strategy of the BBC.

While it is true to say that the BBC's vision for its own local and national services certainly ensured the survival of radio, the role played by the commercial sector must also be celebrated. The initial ILR stations remained committed despite financial uncertainties, and it was the commercial sector's ability to moult its 'independent' skin and embrace a 'commercial' model that in many ways helped ensure the very future of sound broadcasting, not just for itself but for the BBC as well. The true significance of the Heathrow Conference cannot be underestimated here. Street (2006, 209) quite rightly believes it culminated in the 1990 Broadcasting Act which finally allowed the independent sector to become fully liberated to act in its own best interests, and it could be argued it also provoked the BBC into a further realignment of its services and injected a new vitality into the future radio landscape of the 1990s.

The period covered in this chapter was in many ways one of the most dynamic in radio history. A plethora of new ILR stations seeking to establish themselves alongside a BBC desperate to hold its ground, heralded a new competitive arena which forced the actors therein to mould the foundations of a new and more solid broadcasting template. It was an era which also marked a change in the philosophy of radio. The dominant BBC Reithian ideology of the previous decades was initially imposed upon the commercial sector which had previously been immune to its constraints, but it was this very sector which was to develop a system where Reithianism lost its currency and was to be replaced by one where audience maximisation was axiomatic. It was also a period of relative uncertainty. For the commercial sector this came in the form of struggling for financial survival. For the BBC the 1980s was one of the most difficult decades in its history as it came into 'direct conflict with a government deeply suspicious of the very idea of state funded public service and quick to attack the expression of liberal or left of centre views' (Chignell 2011, 151).

The introduction of a legitimate commercial arm did have a number of discernible effects on the UK radio industry. In addition to shifting the core philosophy, it created greater choice and hence a potent competitive element, but that competition was not simply between the BBC and the

ILR stations but rather a circular model of competition where BBC national radio stations, BBC local stations and commercial radio stations were competing with and between each other. In such a turbulent arena, where the commercial sector was attempting to find its foothold while the BBC was endeavouring to regain its own, any spirit of genuine cooperation was lacking. There was perhaps good reason for this. What had been created in this period was in effect a new UK radio industry, one dominated by a long-standing and previously monopolistic player who regarded competition with unease and a new profit-driven player desperately seeking to remodel the very nature of the industry. Chignell (ibid, 139) describes the commercial sector's contribution to journalism as 'unfettered by the caution and smugness of the BBC' and that it 'innovated to the point of recklessness.' This can equally apply to all of the commercial sector's output and therefore any wholesale linking with the BBC was going to be unlikely. As the new industry consolidated, the disparate relationship between the BBC and its new competitors now became firmly cast in stone and would remain so until the necessity of promoting DAB would begin to chip away at the foundations laid down at this time.

REFERENCES

Abramsky, J. "Sound Matters: Soundtrack for the UK." Lecture at Green College, Oxford University, 30 January 2002 (First in a series of four lectures). BBC Press Office, 2002.

Barton, M. (Controller BBC Local Radio). *BBC Radio in the Community*. BBC Lunchtime Lectures, Eleventh Series, 26 October 1976.

Bridson, D.G. *Prospero and Ariel*. London: Victor Gollancz Ltd, 1971.

Briggs, A. *The History of Broadcasting in the United Kingdom. Volume 5: Competition*. Oxford: Oxford University Press, 1995.

Carpenter, H. *The Envy of the World: Fifty Years of the BBC Third Programme and Radio 3*. London: Weidenfeld and Nicolson, 1996.

Chapman, R., 1992. *Selling the Sixties: The Pirates and Pop Music Radio*. London, Routledge.

Chignell, H. *Public Issue Radio: Talks, News and Current Affairs in the Twentieth Century*. Basingstoke: Palgrave Macmillan, 2011.

Corcoran, F. *RTE and the Globalisation of Irish Television*. Bristol: Intellect Books, 2004.

Elmes, S. *And Now on Radio 4*. London: Random House, 2007.

Hendy, D. *Life on Air: A History of Radio 4.* Oxford: Oxford University Press, 2007.

Hill, Lord Hill of Luton (Chairman of the BBC). *Into The Seventies: Some Aspects of Broadcasting in the Next Decade.* Address at Leeds University, 19 March 1969. BBC Publications, 1969.

Jones, S., ed. *Encyclopedia of New Media.* Thousand Oaks: Sage Publications, 2003.

Linfoot, M. "A History of BBC Local Radio in England c1960–1980." PhD Thesis, University of Westminster, 2011.

McCain, T. and Lowe, G. "Localism in Western European Radio Broadcasting: Untangling the Wireless." *Journal of Communication* 40, no 1 (1990).

Peters, K. "Sinking the 'Pirates': Exploring British Strategies of Governance in the North Sea, 1964–1991." *Area* 43, no 3 (2011).

Powell, R. *The Possibilities for Local Radio.* Birmingham: Centre for Contemporary Cultural Studies, Birmingham University, 1965.

Scott, R. (Controller Radio 1 & Radio 2). *Radio 1 and Radio 2.* BBC Lunchtime Lectures, Sixth Series, no. 1 1967.

Seaton, J. *'Pinkoes and Traitors': The BBC and the Nation, 1974–1987.* London: Profile Books, 2015.

Shelley, M. and Winck, M. *Aspects of European Cultural Diversity.* London: Routledge, 1995.

Stoller, T. *Sounds of Your Life: The History of Independent Radio in the UK.* New Barnet: John Libbey, 2010.

Street, S. *A Concise History of British Radio 1922–2002.* Tiverton: Kelly Publications, 2002.

Street, S. *Crossing the Ether: British Public Service Radio and Commercial Competition 1922–1945.* Eastleigh: John Libbey Publishing, 2006.

Trethowan, I. (Managing Director, BBC Radio). *Radio in the Seventies.* BBC Lunchtime Lectures, Eighth Series, no. 4 1970.

Wall, T. "Policy, Pop and the Public: The Discourse of Regulation in British Commercial Radio." *Journal of Radio Studies* 7, no.1 (2000).

Wray, E. "Commercial Radio in Britain Before the 1990s: An Investigation of the Relationship Between Programming and Regulation." PhD Thesis, Bournemouth University, 2009.

Wray, E. *"British Commercial Radio in the 1980s: the Relationship Between Regulation and Programme Content.* Paper presented at 'No Such Thing As Society' Symposium. Bournemouth University, 29 January 2010.

Wright, A. "Local Broadcasting and Local Authority." *Public Administration* 60, no.1 (1982).

CHAPTER 6

Competition on All Fronts: The BBC and Commercial Radio

The 1990 Broadcasting Act was to reform the entire structure of British broadcasting and marked the first step in its total deregulation. It also led to the creation of the first Independent National Radio (INR) stations at a UK-wide level, thus opening up a new arena of competitiveness beyond the local. This ultimately led to a realignment of the BBC services directly affected by this new tranche of competition.

Previously, the main radio battle between the BBC and the legitimate commercial operators was on the level of local broadcasting. However, other unlicensed competitive forces continued to operate. These forces included the ubiquitous Radio Luxembourg as well as other pirate stations such as the enduring Radio Caroline and the new upstart, Atlantic 252. Other variants included the resurgent form of community radio. Many changes were occurring within the BBC too during this time: an expansion of local radio, a restructuring of the national networks, as well as sweeping changes to the running of the organisation as a whole. Television as a form of competition for radio remained quite stable during the 1970s, but by the 1980s its position became at one more engrained and its prevalence spread. Television entered a new dynamic era with the creation of a fourth analogue channel as well as the growth of multi-channel cable and satellite television.

The chapter begins with an analysis of the various non-ILR and non-BBC radio actors before going on to examine the commercial sector itself and finally the BBC in an attempt to highlight the increasingly competitive nature of the UK radio industry. It also examines how the market dynamic

© The Author(s) 2018
JP Devlin, *From Analogue to Digital Radio*,
https://doi.org/10.1007/978-3-319-93070-1_6

impacted upon relations between the actors involved. This forms another backdrop to the uncooperative nature of the relationship between the BBC and the commercial sector which at the end of this period would be transformed, largely due to the impact of technological change.

RADIO FORCES BEYOND THE BBC AND COMMERCIAL RADIO DUOPOLY

Radio Luxembourg

Before the arrival of ILR in the UK, the BBC's most persistent and enduring rival was Radio Luxembourg. It managed to survive the era of the pirates who were as much its competitor as the BBC's and it continued to hold its ground following the creation of BBC Radio 1 in 1967, particularly in the evenings when Radio 1 shared its airtime with Radio 2. So confident was Radio Luxembourg that it cheekily sent the management of Radio 1 a message of congratulations on the latter's opening day (Briggs 1995, 574) and in 1968 Radio Luxembourg decided to take on Radio 1 directly by changing its format from relaying sponsored pre-recorded programmes to broadcasting live with commercial breaks, thus taking the initiative ahead of the ILR stations which would follow some years later. The BBC appeared to have had a better relationship with Radio Luxembourg than its British-based competitors during this period. Following a request from Radio Luxembourg regarding access to BBC audience research findings about Radio Luxembourg and other radio services, the Controller of Radio 1 and Radio 2 was keen to help provide this information,[1] suggesting that Radio Luxembourg may have been seen within the BBC as a different kind of competitor and one that could potentially play a role in keeping the ILR stations at bay.

The confidence that had ensured Radio Luxembourg's survival in the UK over the previous four decades was not to be dented by the arrival of ILR. A year before the new ILR stations came to air, Radio Luxembourg's London Manager Alan Keen was able to say that the new competition:

> ... could not do anything but good and was in fact likely to increase general advertiser and listener awareness of the medium of radio.[2]

[1] Memo from Head of Audience Research, 18 June 1974. BBC WAC: Audience Research, Commercial Radio R78/3, 861/1.
[2] Luxembourg Faces The Commercial Music. *The Times*, 14 March 1972.

Keen's insouciance was based on Radio Luxembourg's performance during the pirate years of the 1960s in which he claimed the company achieved record takings in advertising revenue.[3] Indeed by the early 1970s it was estimated that about 11% of the adult population listened to Radio Luxembourg, rising to 31% in the 15 to 24 age group.[4] Radio Luxembourg had every right not to feel necessarily perturbed by the arrival of ILR for a number of reasons. Firstly, the new stations would be local, whereas Radio Luxembourg covered the entire UK. Secondly, in order to attract advertisers, the ILR stations would be concentrating on boosting their daytime audience which had no effect on Radio Luxembourg as it only broadcast in the evenings, and thirdly, Radio Luxembourg still remained beyond the reach of any needletime restrictions with no British stations, including the BBC, able to match such a density of music output. It was for these reasons that Radio Luxembourg saw its main competitor not as ILR nor even the BBC but commercial television since it was ITV that could attract advertisers and which was 'taking the bulk of the evening audience,'[5] even though by the late 1960s and early 1970s, ITV's advertising revenue started to fall for the first time since its launch.[6] It can also be argued that the dynamic nature of the pirates and the potential of the new ILRs both helped to create a new interest in the medium of radio from which Radio Luxembourg benefitted. For it, competition was always seen as beneficial, but it also tailored its output in a way to cleverly meet the demands of its audience by focusing on a young age group and it was aided in this with developments in transistor technology. These factors had enabled Radio Luxembourg to become a 'cherished part of the UK radio landscape throughout the 1960s' (Radcliffe 2011) and would continue to do so into the 1970s and 1980s.

Radio Luxembourg finally came to an end on medium wave in December 1991, although it continued for a few months thereafter as a satellite channel. Street (2002, 111) suggests that although Radio Luxembourg had managed to survive for so long, it was much diminished in its cultural significance by the age of offshore radio and then the coming

---

[3] ibid.

[4] ibid. It should be noted however that such estimates are not wholly accurate as ascertaining the size of Radio Luxembourg's English audience was difficult to measure (see Worcester and Downham 1978, 701).

[5] Forty Years of Radio Luxembourg. *Daily Telegraph*, 12 August 1974.

[6] In 1969 ITV advertising revenue fell by 1.2% and in 1970 by a further 2.9% (Fletcher 2008, 88).

of Radio 1. This however overlooks the argument that radio's cultural status in general had been shifting downwards since the arrival of television, almost to the point where its position as a cultural form was even becoming discredited among media historians (Hilmes 2002, 5). Street's suggestion also does not take into account the role played by ILR in Radio Luxembourg's eventual downfall. As Radio Luxembourg's English service broadcast pop music only in the evenings,[7] Radio 1 may not be seen as its natural competitor since its evening fayre consisted of more specialist music shows such as those hosted throughout the 1980s by DJs such as John Peel and Kid Jensen.[8] If anything, it was the repositioning of ILR towards the end of the 1980s, playing pop music and transmitting on a superior FM signal, and the prospect of the arrival of Independent National Radio (INR) that sounded the death knell for Radio Luxembourg.

Nevertheless, in historical terms, Radio Luxembourg remains an important forerunner of modern commercial radio in the UK. The satellite and short wave service continued until midnight on 30 December 1992 when the closedown was relayed on various stations, including the old 208 medium wavelength. Thereafter, the 208 service carried an oldies format directed at a German audience. An English-language, classic rock, digital station from RTL Group[9] called Radio Luxembourg emerged in 2005. It was briefly available in the UK using Digital Radio Mondiale (DRM)[10] until 2008. The digital station continues broadcasting over the internet making numerous references to the old 208 service.

Atlantic 252

In 1989, RTL Group teamed up with the Irish state broadcaster, Radio Telefís Éireann (RTÉ), in what the BBC described as a 'curious alliance'[11] to create Atlantic 252, an English-language pop music station broadcasting on long wave from the Republic of Ireland[12] with advertising content

---

[7] In 1989 a daytime English service returned on Radio Luxembourg but aimed at Scandinavian audiences.

[8] Shows like these which broadcast roughly from 8pm to midnight focused on new music.

[9] Radio Luxembourg's parent company.

[10] Digital audio broadcasting technology designed to work over the short wave band.

[11] Board of Management Meeting minutes, 17 Feb 1986. BBC WAC: Commercial Radio, General R92/301/1.

[12] Atlantic 252 broadcast from Trim, County Meath, although it carefully concealed its location from British audiences (Horgan 2001, 154).

specifically directed at the UK audience. Initially the station only broadcast until 7.00pm and ended with an announcement encouraging listeners to switch to Radio Luxembourg for the duration of the evening. Once Radio Luxembourg ceased broadcasting on medium wave in 1992, Atlantic 252 began broadcasting 24 hours a day in order to fill the gap. Atlantic 252 certainly made an impact in the UK, attracting healthy audiences[13] and investment from advertisers, and stayed in operation until 2002 when its long wave frequency was taken over by RTÉ Radio 1.[14] What Atlantic 252 critically proved was that a radio station could still survive on a wave band other than FM. Radio Luxembourg's demise was seen as the result of audiences wanting to listen in better quality, and indeed they could listen to much BBC local radio and ILR on the FM band, but Atlantic 252's success showed that either audiences were content to continue to listen on an inferior band such as long wave or they were not prepared to invest in new sets, whether at home or in the car, that supported FM. As Malm and Wallis (1992, 126) conclude, Atlantic 252 found a good foothold in the UK despite the inherent reception problems of long wave by:

> ... beaming a non-stop pop chart format into western Wales and north-west England after a massive advertising campaign and despite its mono long wave signal. Atlantic 252 has won a considerable following amongst younger listeners in its target area where there is no competition from any local commercial radio station.

Atlantic 252 managed to hold an audience of between three to four million across the UK until 1998 despite not encroaching upon the more populous areas of London and the south east of England. Despite its piratical nature, it embraced many of the general rules that applied to independent radio in the UK and, as Stoller (2010, 217) notes, it generally conducted itself as part of the legitimate commercial community. It was eventually seen off in 2002 by the repositioning of Radio 1 and Radio 2 over the course of the 1990s. Atlantic 252 makes an interesting case study for the student of pirate radio as in many ways a pirate, it also possessed a legal status in that it was licensed by the Irish government and seemed to enjoy a position of privilege compared to other pirate stations.

---

[13] By 1998 Atlantic 252 was achieving and audience share of 1.4% compared to Talk Radio's 1.6% and Virgin 1215's 2.6% (RAJAR Q4 1998).
[14] Irish national network broadcasting a mixture of music and speech programming.

End of the Pirate Renaissance

It could be argued that the arrival of commercial radio in the UK meant that long-established stations such as Radio Luxembourg and newcomers like Atlantic 252 found themselves financially more secure simply because advertisers were now more aware of the benefits of radio advertising and therefore were willing to invest in the medium.[15] Although illegal under the Marine, &c., Broadcasting (Offences) Act 1967, occasional English commercials could be heard on a number of pirate stations such as Radio Caroline throughout the 1970s and 1980s which continued to broadcast from a vessel called *Mi Amigo* and subsequently the *Ross Revenge*, transmitting a 24-hour English service. Another example was Laser 558, launched in May 1984 and broadcasting from the ship *MV Communicator* in international waters in the North Sea. It too attracted an audience within months of setting up, thanks to a strong signal and continuous music fayre, only to close in late 1985. Although this suggests a sort of offshore pirate radio renaissance in the 1980s, the government was determined to put a stop to it and managed to do so once and for all under the 1990 Broadcasting Act. An amendment was made to the Marine, &c., Broadcasting (Offences) Act 1967 within the 1990 Broadcasting Act, known as 'Section 42,' which enabled the British government to take a 'non-territorial or marine approach to offshore radio governance' (Peters 2011). The new clause allowed the government to legally board any radio vessel whose broadcasts reached UK territory.[16] Once 'Section 42' came into effect in January 1991, the last remaining offshore pirate, Radio Caroline, finally succumbed and ceased broadcasting only to reappear for occasional Restricted Service Licence (RSL) broadcasts on satellite and finally via the internet.

Pirate radio's persistence, despite its illegality, may be down to the fact that it had a 'glamorous and adventuresome image for those involved in such activity and its listeners' (Boyd 1986), but one needs to develop a more definitive argument for its ability to sustain a presence. Street (2002, 122) claims much of the pirates' success was to do with broadcasting style and it was the 'reduction in superfluous presentation' by its DJs that had the strongest effect on listenership to both the ILRs and Radio 1. So, we see presentation style coming into play again, spearheaded by the pirates as they had done in the 1960s. Laser 558, for

---

[15] Commercial Radio: the 19 stations that showed why the cynics were wrong. *Campaign Magazine*, 2 July 1976.

[16] This was also now permitted under the UN Convention on the Law of the Sea.

example, used mostly DJs from the USA who employed 'snappy techniques but with the emphasis very much on the music' (Skues and Kindred 2014). Similarly, Atlantic 252 was playing 'long sweeps of music with minimal input from DJs' (Taylor 2003, 73). Once again it was radio stations broadcasting from beyond the UK's regulatory boundaries which were leading changes to radio broadcasting which would eventually permeate the entire industry. By comparison, Radio 1 may have seemed too conservative while the ILRs were constrained by excessive regulation, thus leaving a powerful vacuum to be filled by a new wave of pirate stations. The BBC remained concerned by the continued existence of the pirates, identifying three still broadcasting to London by the late 1970s and all on the same FM band at different times of the day (London FM, Radio Invicta and Free Radio London).[17] Indeed the sheer gall of the pirates was sometimes hard to comprehend:

> Radio Caroline is still broadcasting in contravention of the Marine Offences Act. It has a mobile disco roadshow playing at halls in the south east of England. They advertise their appearances with apparent impunity.[18]

The Urban Pirates

By the late 1980s a new breed of pirates had begun to operate, not on the high seas but 'cloaked in the anonymity of urban sprawl' (Mason 2008, 43). The new pirates operated wholly on the clearer FM band, catering for a new generation of radio listeners particularly in London. These pirates were more difficult to tackle. The authorities can detect a pirate's homemade antenna easily, usually tacked to the top of a tower block, but the studio connected to the antenna by a less powerful and undetectable microwave signal is more difficult to track down. Transmitters were found and confiscated, but studios were harder to find and stations selling advertising could afford to replace lost antennae within hours and thus began 'a game of cat and mouse with the authorities' (Mason ibid). How these new urban pirates differed from the previous pirate model was that they were not catering for a general pop music audience but instead were providing

[17] Memo from Chief Assistant, Radio Management, Programmes to Deputy MD Radio, 30 August 1978. BBC WAC: Commercial Radio, General R92/301/1.

[18] Letter from Douglas Muggeridge, Deputy MD Radio to Broadcasting Department, Home Office, 1 September 1978. BBC WAC: Commercial Radio, General R92/301/1.

airtime for various musical sub-cultures and with so much success that Mason (ibid) claims by 1997 the specialist pirate stations in London were attracting as much as 10% of the capital's radio audience. Urban musical genres such as hardcore, techno, drum 'n' bass, UK garage, grime and dubstep are some underground movements that developed with the help of these stations, and it all began with the Acid House explosion of the late 1980s (Collin 2009), a movement that unlike the mainstream music industry was characterised by 'decentralisation and independent productions' (Hesmondhalgh 1998) which would in themselves find little outlet on mainstream radio.[19]

A fine example of a station demonstrating this new pirate broadcasting model was Kiss FM which began in 1985, broadcasting hip-hop and house from the suburb of Crystal Palace in south London (Goddard 2011) with a roster of DJs that included Tim Westwood, Pete Tong, Trevor Nelson and Dave Pearce, all of whom would eventually move to Radio 1 when it decided to cater for this very audience. By 1990, Kiss FM was so popular, it was granted a licence and went on to become a multi-million-pound media franchise turning over £161 million in 2005 (Mason 2008, 45) with the executives in charge still recognising where its kudos came from and still recruiting pirate DJs from other underground stations to replace the bigger names once they had moved to the BBC or the big commercial music stations. This represents an interesting pattern of progression for broadcasters which is similar to that which occurred when Radio 1 came to air in 1967. DJs establish their reputation at pirate stations and on achieving a certain status are then recruited by the BBC to deliver the same output but to a national audience. In many ways this works to the BBC's advantage, as it can survey the underground radio landscape to see what listeners really want to hear then hire established voices to attract that very audience to Radio 1. It would appear that this might be a parasitic relationship with the BBC engorged with an audience initially fostered by the pirate stations, but it also shows that the role of the pirates had changed dramatically from being a provider of populist music to being at the cutting edge of new musical genres. Ultimately, it places the power in the hands of listeners as it was, for example, their support for Kiss FM which led the authorities to grant the station a licence and ultimately

---

[19] Radio 1's sole contribution to the dance music scene at the time consisted of a three hour show on Friday nights, *Jeff Young's Big Beat Show*, until a reorganisation of the network in the early 1990s brought in programmes catering for other genres.

led the BBC to provide similar content which would cater for their tastes. Perhaps the BBC's position had mellowed too and its attitudes to the pirates had changed from the damning indictment of the 1960s:

> The pirates were efficient thieves. They stole wavelengths. They stole news. They stole the copyright of composers, musicians and disc makers. (Edwards 1968).

Community Radio

Community radio had come about from 1983 with the formation of the Community Radio Association (CRA). This resulted from an increasing number of pirate transmissions which emerged from within ethnic groups feeling disenfranchised by existing media, and McCain and Lowe (1990) argue this strand of radio was 'fueled by pirate radio stations' successful challenge to state-run services.' In 1985 the Home Office set up an experiment in community radio as evidence suggested that there was a high level of interest for 'community of interest and neighbourhood stations' (Lewis 1985). This experiment was short-lived as it was abandoned a year later by the government, leading those licence applicants to vent their frustration through protest in order 'to tell Londoners how they had been cheated of community radio' (Goddard 2011, 48). Despite the Conservative government's initial belief that community stations should have a role and should exist as part of a broader spectrum of radio, they failed at this point as a broadcasting model, and Scifo (2011) believes this was because of government concerns about 'the consequences of non-balanced broadcasting.'

The notion of community radio is a useful one to attempt to define as it was a term that had been bandied about for some time and pinned to the mast of many different radio campaigns including BBC local radio and ILR. It is probably best summed up in Stoller's (2010, 154) succinct definition of what is often referred to as the third tier of radio in the UK:

> It is a local service, run largely on a not-for-profit basis, involving a significant volunteer element and concerned to deliver some social benefits beyond just the provision of an on air radio service. The station also needs to be independent in the sense that it is not controlled by or beholden to mainstream media and in most cases will be stand alone.

Stoller (ibid, 155) also discovered the first documented use of the term Community Radio in a paper by Rachel Powell of Birmingham University in

1965. Powell (1965) argued for up to 250 not-for-profit local stations financed by local government and the BBC licence fee. Stoller (ibid) also argues by the time of the discussions around the 1971 White Paper[20] on the future of radio, the tone of the community lobby was overwhelmingly one of doctrinal hostility to ILR. This is interesting in that it exposes an important fact, namely that a vociferous community radio lobby co-existed alongside a commercial radio lobby with the purpose of similarly catering for a niche audience and providing popular content, differing only in their attitudes to profit.

Naturally the IBA was concerned as it assumed the existing ILR system was already providing an 'effective and self-financing form of local community radio'[21] and it made great efforts to provide examples of how ILR stations were 'contributing to community concerns.'[22] By the late 1980s, the regulators had responded and thereafter community radio re-emerged and took one of three new forms: new 'incremental' stations[23] launched by the IBA in 1989, short-term Restricted Service Licences (RSL) developed by the Radio Authority in 1991 and small-scale ILR licences deployed on the expanded FM spectrum after 1994. The 1990 Broadcasting Act placed no specific community radio obligations on the new regulator—the Radio Authority—and it looked at that point as if we might witness the demise of the third tier of radio in the UK. The commercial radio sector remained opposed to the notion of a separate community radio strand (Gordon 2009, 68) and indeed political support had waned. The CRA eventually lost heart too and renamed itself the Community Media Association (CMA) and began to pursue the field of local television. By 1998 the Radio Authority had become more sympathetic to the development of a community radio sector. Following the Communications Act 2003 and the Community Radio Order 2004, the new communications regulator Ofcom[24] began to licence community radio stations. By April 2008 around 170 licences had been issued. With this, the definition has once again changed, now such stations must achieve some degree of 'social gain' which, according to Ofcom, will be achieved when:

> … members of the community that the station targets have shown achievement… when the stations broadcast to those who are

[20] White Paper: An Alternative Service of Radio Broadcasting, 1971 (Cmnd 4636).
[21] IBA Press Statement, 25 July 1985.
[22] Independent Local Radio: Serving the Community. IBA, December 1977.
[23] Stations with a community remit but operating within an existing ILR space.
[24] Office of Communications, formed in December 2003.

underserved... when they facilitate discussion, provide training for volunteers... lastly help towards a better understanding of the community.[25]

The third tier therefore re-emerged in a new and more plural radio environment and one which, one could argue, has been unwittingly created by both the BBC and the commercial radio sector who, by the early 1990s, had diminished once and for all the threat from the pirate sector—with the help of government legislation—leaving the radio arena, with the exception of small urban pirate concerns, open to a battle between these two radio behemoths.

TELEVISION

Since the arrival of ITV in 1955 the television landscape had remained relatively stable compared to that of radio, the only major developments being the launch of BBC 2 in 1964 and a number of rounds of ITV franchise changes (Johnson and Turnock 2005, 15). But the three existing channels were to face a new rival following the publication of the Annan Report in February 1977[26] which was to mark a profound change in the UK's television ecology. The Annan Committee had recommended a fourth television channel, but in rejecting ITV's call for an ITV 2, it proposed instead an independent network. The result was Channel 4 and it came to air on 2 November 1982. Channel 4 was created by the 1980 Broadcasting Act and it was to be a publisher of programmes rather than producing in-house. It was to be a public service broadcaster which, although commercially funded through advertising revenue, would remain publicly owned, thus representing an entirely new model of broadcasting. Its unique nature was also cemented by the requirement to be innovative and cater for minority interests not found on the other channels.[27] While rumblings were beginning to take place in the radio industry regarding the constrictive nature of Public Service Broadcasting (PSB), the opposite seemed to have been taking place in television as experimentation and minority programming were the cornerstone of Channel 4, marking an explicit extension of the public service ethos.

[25] Community Radio Order 2004. London, HMSO.
[26] Report of the Committee on the Future of Broadcasting (Annan Report), 24 February 1977 (Cmnd 6753).
[27] See Brown (2007) and Hobson (2008) for more in-depth studies of Channel 4.

The BBC, while unhappy with the notion of a fourth channel from when the idea first raised its head in the early 1970s, did nevertheless support the idea of a network catering for minority audiences and ultimately one not run by ITV:

> The IBA is advancing a case for the allocation of ITV 2... they say they wish to cater for a number of interests which they cannot serve as adequately as they would wish. So what does the BBC propose? We say there is no case for a fourth general television network but there is strong case for allocation of a fourth network to genuine minority viewing (Curran, 1974).

Interestingly at this time the BBC Director General, Charles Curran, believed that a new channel would not actually increase the total amount of television viewing, claiming this did not occur in 1955 with the launch of ITV nor in 1964 with the launch of BBC 2 (Curran ibid), thus suggesting the BBC had little to fear from a new channel affecting either its television service or its radio service.

Another hugely significant development took place in television only a few months later and that was the arrival of breakfast television which many believed would have a 'more definitive impact on radio.'[28] The BBC's contribution known as *Breakfast Time* went on air on 17 January 1983, followed by the winner of the ITV franchise, *TV-am*, two weeks later on 1 February. Although Channel 4 appeared to have no direct impact on radio other than being yet another televisual distraction which the radio industry had now accepted as a norm, breakfast television was another matter. ILR companies particularly became nervous about the prospect of direct competition from breakfast television due to the fact that this is the exact time of day when radio audiences are traditionally at their highest and it is generally considered to be the main entry point to a station as well as an attractive point of investment for advertisers and record companies (Garner 1990). Breakfast television's threat potential was readily apparent to the ILR stations and the possible impact it might have on their financial viability, so much so that the AIRC had argued that any introduction of breakfast television should be delayed until ILR coverage reached 85–90% of the UK, which would not be expected until October 1984 (Stoller 2010, 129). Breakfast television was

---

[28] Breakfast TV: The Impact on Radio. BBC Audience Research, July 1983. BBC WAC: CS/83/06.

to be a whole new experience for British audiences and they took to it keenly, much to the detriment of radio. In fact by the end of 1983, it had caused a 10% drop in the amount of time per week which the average person spent listening to the radio (Crisell 1994, 35). As Wieten and Panitti (2005) note, unlike other day parts, breakfast television 'identifies closely with the viewer, consciously adopting the role of structuring its schedule to meet the interests of the morning audience.' It would be reasonable to assume that those in radio, both BBC and ILR, would have been concerned about breakfast television and how it might displace radio listening at that important time of day. Street (2002, 124) claims the creation of new television companies and particularly breakfast television had an 'indirect, although major impact on independent radio' (although I could argue this applied to all radio) and 'struck the industry at almost its weakest moment,' reducing the breakfast audience for radio by 10% over the course of 1983/1984. Indeed by 1985, one in five radio listeners said their radio listening had been reduced by watching television at breakfast time. While most of these claimed their early morning radio listening had been substantially reduced, few said they had completely abandoned radio at this time. Only 15% who said that breakfast television had changed their listening habits had stopped completely, and they represented less than 1 in 20 of all radio listeners (Gunter 1989, 80). This suggests that radio listening at breakfast had declined to some extent as a result of breakfast television, but the decline had probably come to an end with most radio listeners' listening patterns remaining largely intact, and by 1989 the crisis seemed to have dissipated with the total breakfast television audience having shrunk to 3.6 million in any one day as the similar radio audience figures recovered (Garner 1990).

The 1980s also saw the arrival of cable and satellite television services in Europe. This represented a new arm in the British television market. The years of BBC/ITV rivalry now seemed antiquated as first Channel 4 joined in, followed by the cable and satellite services. The 1984 Cable and Broadcasting Act led to the introduction of cable TV to the UK after franchises were awarded by the Cable Authority which was created in 1985. Direct satellite broadcasting began in 1989 by Sky TV followed by British Satellite Broadcasting (BSB) in 1990.[29] Cable TV gave viewers access to a different form of television for the first time, namely, one not provided by the old BBC/ITV duopoly and therefore not constrained by any broad

[29] Both companies merged in 1991 to form BSkyB.

public service remit. Likewise, satellite TV offered a similarly attractive and unconstrained amount of programming. It had to of course as viewers were required to pay for it via subscription—a novel method for consuming media in the UK. This new element of freedom in television broadcasting came about around the same time as similar steps were being made in the radio domain. Crisell (2002, 232) believes in fact that the cable and satellite operators sought to 'follow the lead of radio and maximize their audiences' not simply by *narrowcasting* but by *formatting*, that is, by providing specialised audiences with only one type of programming on different channels, such as specific movie, comedy or sport channels. Crisell is indeed correct in this. The rumblings taking place in radio since the 1984 Heathrow Conference were markers of a potential break-up of the existing broadcasting regime based on a strong PSB remit. This, coupled with the political and economic climate created by an enduring Conservative administration, meant the possibilities of a freer broadcasting environment were likely to be imminent and potential radio and television entrepreneurs were, unsurprisingly, poised.

COMMERCIAL RADIO: ACHIEVING A NATIONAL PLATFORM

The 1980s was for commercial radio a decade of struggle; financial struggle and a struggle to reposition and redefine itself. It was also a decade of great expansion for the sector. In the first year of the decade, 26 stations were broadcasting to a weekly adult audience of over 14 million, attracting advertising revenues of over £44 million. By 1990, 79 stations were reaching an audience of over 22 million with advertising income reaching almost £125 million. The early years of the 1980s were difficult for many stations with the exception of Capital Radio in London which even in 1982 could claim to be both the richest station in the country and the number one station in London.[30] The example of Capital was good for the rest of the industry as by the mid-1980s ILR was emerging from the recession which had been gripping all sectors of the economy. By September 1986 two thirds of the 40 ILRs reported profits with Capital Radio increasing pre-tax profits by 80% to £1.7 million (Stoller 2010, 165). Indeed the radio sector's total market value was to rise from £33 million in 1987 to £400 million by 1990 (Stoller ibid, 200). Alas, the story was not common to all stations. In October 1983 Centre Radio based in

---

[30] Capital Radio – In Tune With London. *Financial Weekly*, 13 August 1982

Leicester was the first commercial station to fail as it went into liquidation only two years after launch.[31] By the late 1980s, there was a feeling that the tough days were over, but the approaching deregulation that had originally excited so many in the commercial sector was now seen by some as a 'velvet-clad fist.'[32]

A crucial development within the commercial sector at this time were the 'incremental stations.' Former pirate station, Kiss FM, was one of 20 incremental stations set up during 1989/1990 as an experiment to provide localised and specialised programming to serve minority audiences which were not catered for by the BBC or ILR.[33] It was the IBA which came up with the idea of incremental stations by proposing new services confined within existing ILR areas that allowed for a more specialised or more local service but still within the requirements of the 1972 Sound Broadcasting Act. Since they were additional services, they came under the nomenclature 'incremental stations.' Often with little commercial experience, the incremental stations, though distinctive, found themselves competing with BBC local services as well as the established ILRs in the same terrain. Suddenly competition from within the commercial sector itself seemed less certain to bring benefits. Capital was to face music radio rivals in London in the form of Kiss, Melody and Jazz. Some of the incremental stations of course remained true to their roots, London Greek Radio for example, and maintained their position as a genuine form of community radio, but others quickly evolved into commercial stations in their own right. This despite the IBA having been explicit when announcing its plans that these stations should be clearly 'community of interest' stations.[34] It is important to note that many of the incremental stations ultimately failed due to the forces of market economics and niche programming and that those that were able to survive did so precisely because they ignored this directive from the IBA. By 1992, 25 such licences had been advertised, 2 of which eventually failed completely and fell off air. Of the remaining 23, few apart from the ethnic stations in high population areas survived with their original remit intact.

Some commentators attacked the existing promises of performance which limited the established ILRs, while incremental stations appeared to

[31] Commercial Radio: Profits and Problems in the Air. *Financial Times*, 5 November 1983.
[32] More Competition Gives Greater Urgency to Radio's Image Problems. *Campaign Magazine*, 4 November 1988.
[33] Other examples in London included Melody Radio and Jazz FM.
[34] IBA News Release, 9 November 1988.

be able to shift their output as they came under economic pressure.[35] So, for example, some ethnic stations began playing mainstream music in order to attract listeners outside what had been considered their original core audience. This period is therefore characterised by an onslaught on ILR not from external forces but from within, namely, from the IBA which was responsible for creating yet another layer of competition within the existing commercial sector. Beyond the competitive arena of London, where Capital and LBC constantly battled against each other over audiences, many ILR stations were now to face direct competition within their own franchise areas. Despite this, the already hugely successful mainstream 'heritage' ILR stations remained dominant in the radio industry. Capital Radio plc made a profit of nearly £16 million in 1990 despite the economy at large being in the throes of recession.

Regional licences represented another step in introducing competition for the 'heritage' ILR stations. The Radio Authority announced in September 1992 its intention to advertise five regional licences to come on air in 1994. These would encompass the territory of two or three conventional ILR services and would be the largest coverage licences outside Greater London. Stoller (2010, 230) identifies a number of effects of these regional licences on the existing commercial sector:

> They took the ground from under the feet of those in the 'heritage' ILRs who might have maintained voluntarily the old public service model of independent radio for longer... Also, losing their monopoly and facing new commercial competition they (the 'heritage' ILRs) had to cope with the freeing up of the radio market which they had themselves set in train at Heathrow.

Stoller's idea that some of the existing ILR stations would have been in a position to protect the public service dimension of the commercial sector may be true to a point, but the key words he uses are 'for longer.' There is certainly no suggestion that it was a value that could have been permanently maintained in the very long term. A momentum had been gathering steam since Heathrow and unbridled commercialism was now a force which seemed unstoppable.

The 1987 Green Paper, Radio: Choices and Opportunities,[36] was the response to the challenges issued by ILRs at Heathrow and thereafter. It

---

[35] Terry Smith, MD Radio City. *Media Week*, 26 January 1990.
[36] Radio: Choices and Opportunities. Green Paper, Cm 92, London: HMSO, February 1987.

recognised changes were needed in the regulatory framework for ILR. The paper paved the way for the 1990 Broadcasting Act, thus heralding a number of changes critical to the history of British radio. The core changes of the Act were; a demarcation from television by the creation of a separate regulator, namely the Radio Authority;[37] the loosening of the public service remit; permitting consolidation of ownership of radio companies; and the launch of Independent National Radio (INR).[38] For the ILRs it can be argued the regulatory processes changed overnight under the Radio Authority as the Broadcasting Act 1990 radically shook the UK's public service ecology:

> Prior to 1990 all broadcasters were public service broadcasters...
> after 1990... broadcasters were only required to abide by negative content regulation... and not positive content obligations (Cowling 2004, 64).

A new found financial confidence led to the creation of media companies attracted to profit who saw a healthy return could be achieved from sound broadcasting. This new found confidence, which took hold in the latter part of the 1980s, meant the ILRs could continue their protest against the IBA's restrictive nature and direct independent radio even more down the commercial path. The Peacock Committee, which reported on 29 May 1986,[39] had bolstered the ILRs' stance by proposing that IBA regulation of radio should be replaced by a looser regime. Bizarrely, Peacock also proposed the BBC should take over failing ILR stations or indeed successful ILRs could buy out a BBC local radio station if the BBC was willing to sell, thus envisaging a scenario which never came to fruition of full blown competition between the BBC and the commercial sector essentially fighting over each other's assets.

As commercial prospects improved, so did the influence of larger radio groups who became the new players on the commercial radio stage as the rules of ownership changed. The 1990 Broadcasting Act limited companies to owning no more than 20 local stations and of those, no more than 6 could be large licences. This accumulation of ownership saw the emergence of some of the companies which would in the future become the

[37] Formed on 1 January 1990.

[38] The Act also catered for the expansion of Community Radio although these plans remained undeveloped until well into the 2000s.

[39] Report of the Committee on Financing the BBC (Peacock Report) 1986 (Cmnd 9824).

major players in leading the commercial sector. So by 1994, companies such as Capital, Emap, Scottish Radio Holdings, TransWorld, GWR and Chrysalis had made their mark on commercial radio. In fact these companies began to demand even more freedom to expand, and this happened under the later 1996 Broadcasting Act which permitted even greater concentration of ownership so that there were 14 separate mergers or acquisitions in 1996 and a further 13 in 1997.

With investment and expansion these companies brought some important changes to programming philosophy within the commercial sector, which it can be argued truly secured the sector's survival. Not only that but they also ensured its very dominance over the BBC at this time. In July 1995 the Radio Authority received a request from the GWR group to be allowed to network programmes across virtually all its stations, that is, a dozen FM stations and eight AM stations. It was aiming to transform its disparate portfolio of local stations into a network and as Stoller (ibid, 255) notes, Steve Orchard, the Head of Programmes at GWR wanted to:

> ... improve the quality of presentation for listeners in local areas and to help even our smallest sites to compete against BBC national networks... We are now actively negotiating with a number of national radio personalities who are ready to 'come over' to the commercial side.

Thus we see for the first time the commercial sector setting out to poach BBC talent in a complete reversal of what had happened with the launch of Radio 1 in 1967 or when the urban pirates appeared in the late 1980s when the BBC was poacher. Critics could however argue that the names moving from the BBC to the commercial sector at this point were those whose own BBC careers may have been coming to an end anyway.

The arrival of the big radio groups in the early 1990s represents a landmark in British commercial radio history simply because they represent a new found economic confidence in the medium, but also because they gave the commercial sector the ability to compete on a par with the BBC for the first time in its history. It also shifted the balance of power in the commercial radio industry from the regulator (the Radio Authority) to the largest commercial companies. Any notion of independent radio was now totally erased by a genuine commercial drive in the radio broadcasting market. By 1993 advertising income had jumped from £141 million to a record £178.3 million and share prices soared (Stoller ibid, 245). Stoller goes on to say that these were years of personal and corporate money making but often at the expense of investment in programme output which is an interesting dichotomy but surely an unsurprising one as a pure commercial model could only cater for a populist output, and to be free of the shackles of public service duty was what the sector had been desperate to achieve for a number of years.

Another aid to this was the proposal to abolish simulcasting, that is, broadcasting the same services on FM and AM wavelengths. By the end of 1989, split services were now provided by many stations, in effect giving the existing commercial stations the opportunity to target two separate audiences. Thus we see Capital offering a golden oldie service *Capital Gold* on AM which had the effect of actually increasing Capital's share of London's listening from 17% to 28%. LBC's split into separate news and talk stations was predicted by its owners, Crown Communications, to increase its listenership by 15% as its news station (LBC Crown FM) and its talk station (Talkback Radio) directed themselves towards Radio 4 and Radio 2 respectively.[40] However by 1992 it became apparent that the investment required to operate separate networks was having a detrimental effect on return as Capital's profits went down by a third over 1991 with the same happening to LBC. The split between FM and AM in 1989 had not worked with both stations losing some audience (Stoller ibid, 227).

It is critical to note at this point the important role played by the commercial sector in radicalising content. From 1987, Capital's new current hits radio format began to change the face of pop music radio in the UK and its impact on the BBC was profound. This change of content became nominally known as Contemporary Hit Radio (CHR) which can be described as:

> A rock/pop music format that plays the current best-selling records.
> The music is characterised as lively upbeat hits. The playlist generally
> consists of 20–40 songs played continuously throughout the day. Disc
> jockeys are often upbeat personalities and the format emphasizes contests
> and promotions… targeting a young demographic of both men and women
> aged 18–34 with a listenership extending into the 35–44 demographic cell
> (Lee 2004, 621).

It was introduced at Capital Radio by Richard Park, spread throughout the ILR system, and Stoller (ibid, 164) believes it gave ILR almost ten

---

[40] Crown Promises LBC Listening Boom After Split. *Broadcast Magazine*, 2 September 1989.

years' dominance of the centre ground of popular music radio, much to the consternation of the BBC.

Another symbol of success occurred in July 1993 when commercial radio launched a new network chart show sponsored by Pepsi called *The Pepsi Chart Show*. At different stages 80–100 ILR stations carried the networked show. It became the most listened to radio show in the UK with an audience of 3.6 million at its peak, almost 1 million more than the equivalent Radio 1 Top 40 chart show. Yet again it was not the BBC but its competitors who were taking the lead in remoulding British radio. Where the competition took the lead, the BBC simply seemed to follow. If this could happen at a local level, then surely the time was right for it to happen at a national level too.

The most striking proposal of the 1990 Broadcasting Act was that of Independent National Radio (INR), a term that although lurking in the radio background for some time, became the new radio buzzword. The notion of national commercial radio was not a new one. Stoller (ibid, 171) claims in the debates around offshore pirates in 1960s there was a moment when the BBC's 247 medium wavelength might have been put aside for a single national commercial pop music channel. Later in 1976 the IBA was reflecting on whether it should seek a national channel as at this point it believed:

> Independent Radio can only be a full, credible and effective alternative to total BBC radio (as against local BBC radio) if it has both local and national channels.[41]

By 1980 the IBA was drawing up preliminary plans for a national commercial station with the aim of breaking the BBC monopoly in UK-wide national radio. As the IBA's head of radio John Thompson stated:

> The IBA has made it known to the government that in the right circumstances the authority might wish to see this final, lingering state broadcasting monopoly challenged.[42]

The use of words like 'in the right circumstances' and 'might wish to see' do however reflect the fact that at this stage a national network was not the primary driving force for the independent sector, instead the urgent priority was the continued development of ILR.

[41] IBA Paper 194 (76), 19 July 1976.
[42] National Radio Rival to BBC Planned by IBA. *Daily Telegraph*, 20 November 1980.

A concerted lobbying effort from commercial broadcasters persuaded Home Secretary Willie Whitelaw to announce in March 1983 that the UK would seek international approval for the allocation of VHF frequencies for a new independent national radio network along with a further BBC service on VHF.[43] By 1986 the IBA was able to boast that it could have a national radio service on air within four years. John Thompson's message was now less equivocal:

> The BBC still enjoys a total monopoly in national radio… this is the only monopoly in broadcasting still to survive. The BBC's tenure as the sole supplier of national radio needs in the public interest to be challenged.[44]

The key issue for a new national commercial network was going to be centred around the question of spectrum. For a new network to have a national presence at this time it would have to be allocated an existing BBC medium wave position. By 1989 the BBC was made to give up 1053 and 1089 kHz for one new INR service and 1215 kHz for the other, and Stoller (2010, 172) describes a final decisive meeting between the BBC and IBA at the Home Office in 1989 where the BBC, under pressure from the government to offer spectrum to the commercial sector, argued that 'the commercial sector would only need medium wave frequencies, since audience figures showed nobody listened to radio after six o'clock at night' to which Douglas Hurd, the Home Secretary, agreed, thus leaving the commercial sector with the impression that any chance of a national network on FM was now lost.

As the Broadcasting Bill went through Parliament however, it became apparent that an FM frequency would be employed alongside two AM licences. Fears that the three new stations would simply provide near identical pop music were to be assuaged by the government who guaranteed the three new services would each be devoted to a different theme and the first national commercial station would provide classical music on FM while the AM services would be for news and music.[45] It would appear that the commercial sector's recent experience of what one can call 'intrasector competition,' that is, of competing within itself, had now made it reluctant to support a national pop music network, in other words, a type of network that had become the mainstay of the sector. There was therefore

[43] New Frequencies: Same Radio Shows. *New Scientist*, 21 April 1983.
[44] BBC Radio Rival 'Inside Four Years.' *The Times*, 24 February 1986.
[45] Thatcher Orders More Specialised Radio. *Financial Times*, 17 April 1990.

relief within the Radio Authority when the government announced on the 30 October 1990 that the FM licence would be for a 'non-pop music service,'[46] thus allaying fears among the ILR stations that the biggest threat to their existence would come not from the BBC but from other commercial companies willing to take the profit motive to a national level.

When the Radio Authority advertised the first national licence in the national press on 11 January 1991, there was a feeling that at least the new national networks were more likely to complement rather than compete with ILR stations. Classic FM was subsequently awarded the first national commercial licence and began broadcasting on 7 September 1992. As a station devoted to classical music, Classic FM's achievements were cited in comparison to BBC Radio 3 which also concentrated largely on classical music and indeed from the outset the former performed well, attracting a weekly audience of 4.3 million after only five months on air, while Radio 3's audience dropped to 2.5 million, making Classic FM now the fourth largest radio station in the country.[47] Its success against Radio 3 was largely due to the fact that it delivered classical music in a different format—populist, shorter, light classical music with a familiar presentation style—and successfully marketed an output that previously appealed only to minority tastes. Although in terms of content, Classic FM's natural competitor may have been envisaged as Radio 3, its different format and style of presentation meant that its actual rivals were just as likely to be Radio 2 and Radio 4. Thus we see the commercial model at the national level employing its skill at adapting format and style to encroach upon the BBC's national services.

On 2 April 1992, the Radio Authority awarded its second national licence to Independent Music Radio, a consortium owned by TV-am plc and Virgin Communications Ltd. The station was to be called Virgin 1215,[48] and it began broadcasting on 30 April 1993 with an adult-orientated rock format and with the backing of Richard Branson.[49] Like Classic FM, many would have envisaged Virgin to have a natural BBC competitor, in this case Radio 1, but like Classic FM it also traversed the single competitor boundary and both Radio 1 and Radio 2 began to feel its effects. It too built an audience managing to quickly achieve early,

---

[46] Radio Authority Press Release, 30 October 1990.
[47] RAJAR Q1 1993.
[48] 1215 referring to the medium wave frequency.
[49] Founder of Virgin Group.

respectable weekly figures of 2.16 million by autumn 1993.[50] That figure was to rise to almost 4 million by the end of 1994[51] but eventually settled and did not exceed 3.5 million for the remainder of the 1990s, meaning it was on occasions outflanked by Atlantic 252 and the BBC's poorest performing station, Radio 3.[52] Virgin management were all too aware of the problem, namely that it could only be heard on medium wave and the recent demise of Radio Luxembourg on that wavelength could have done little to boost their confidence. With reception in many parts of the country, including the all-important London and the south east area, 'patchy or even non-existent,'[53] Virgin's management knew it had to find a presence on FM. This was of course a desire even before the station launched, and Stoller (2010, 217) reveals a meeting between the BBC and the commercial sector when, just a few days before launch, Richard Branson visited BBC Director General, John Birt, to discuss the possibility of swapping frequencies, with Branson swapping the 1215 medium wave frequency for Radio 4's 93.5 FM frequency. Branson however came armed with a threat; should Birt relinquish the frequency, then Branson would save him having to bid for Radio 1 when it was privatised—something which was still a possibility at the time. In the end the Radio Authority eventually offered Virgin an FM licence for the London region from April 2005.

The final national network was to be a speech network and Talk Radio[54] began on 14 February 1995. It was a speech network certainly not modelled on Radio 4 but instead consisted of 'shock jocks' with the unadulterated objective of provoking the audience. This was a style of broadcasting which, while familiar in the USA, had not been very widespread in the UK except for some late night talk shows on a few ILR stations but now was to have a national audience. The verbosity often got Talk Radio in trouble however and after incurring numerous fines it had to tone down its output.

The early years of the 1990s are hugely significant for the commercial sector. It had thrown off the shackles of the IBA and was now operating in an environment it had been striving to create since the Heathrow

---

[50] RAJAR Q3 1993.
[51] RAJAR Q4 1994.
[52] Atlantic 252 average audience = 3.9 million, Radio 3 = 3.1 million. (Source: RAJAR).
[53] *Daily Telegraph*, 18 November 1993.
[54] Now known as Talksport.

Conference. Stoller (2010, 244) says this period was a great time to be in private radio in the UK:

> It still retained characteristics of independent radio to make it seem a worthwhile undertaking judged from a broad social perspective while at the same time enjoying the fruits of commercial success.

It could be argued that any remnants of independent radio had now totally disappeared and a commercial ethos permeated the entire sector, thus begging the question if any of the commercial companies were genuinely maintaining a social responsibility at this point. If Stoller marries the notion of independent radio with a public service responsibility or indeed a nod to Reithianism, then his statement may be inaccurate as the critical shift that occurred in the commercial sector at this time rendered it totally detached from the attributes of the erstwhile independent radio model. The expansion of ILR,[55] the arrival of networked stations and finally the holy grail of national stations meant that a paradigm shift had occurred in the independent sector, that is, it had now become a fully fledged commercial sector. This was also reflected in its performance. In 1993, the 20th birthday of the commercial sector, audience numbers were up by two million for all commercial radio and advertising revenue was recovering as the recent recession came to an end. It now had 3 national stations and 170 local stations offering 30,000 hours of programming per week with an audience of 26.4 million adults and taking 43% of all radio listening in the UK (Stoller 2010, 234). Commercial radio had seemingly achieved its goal.

BBC RADIO: A PERIOD OF CRISIS

The BBC's attitude towards the independent radio sector in the latter's early years was, to say the least, relaxed. By the end of the first year of independent radio, the Controller of Radio 1 and Radio 2, Douglas Muggeridge, was not unduly concerned:

> Today as we approach the first anniversary of commercial radio I am able to report BBC radio has experienced no loss of audience, if anything our figures are running slightly higher than last year. The large audiences quoted

[55] Sixteen new ILR licences were awarded in 1992.

by the commercial stations bear no relation to our figures.[56] It should be remembered that the 6 commercial stations cover nearly 40% of the population or 20 million people per week, this is not much less than our own BBC local stations which can be received by approximately 25 million people per week.[57]

What permeates the early years of the independent sector is a scant recognition of its position as any meagre form of threat from within the BBC. A perusal of the memoirs of former Directors General Charles Curran (Curran 1979)[58] and Alasdair Milne (Milne 1988)[59] proves this point as they both refer only fleetingly to the independent radio sector in what are otherwise extensive reminiscences from the world of broadcasting. It could be argued that for the BBC, localism in radio had diminished in importance from its heyday in the late 1960s and came lower down the pecking order of priorities even as far as its own services were concerned, and this was backed up by a future Director General, Ian Trethowan:[60]

> Local radio has been a factor in binding together an area but whatever has been achieved by different forms of community radio and however it may be developed of the next decade, in the foreseeable future the heart of BBC radio will be the national networks. It has been through the networks that we have been able to offer listeners not only a wide range of programmes but programmes which have aimed at the highest quality. It's where we sustain our cultural output and daily journalism (Trethowan 1975).

If any perceived threat from the commercial sector went unappreciated during the 1970s and 1980s, it certainly was to become a reality towards the end of the 1980s as the commercial sector both expanded in range and experimented in output. While the new commercially driven local stations were in themselves a major concern, particularly for Radio 1, it was the

---

[56] Disagreement between the commercial sector and the BBC over audience figures and the methodology employed to calculate them would persist until a single measurement system agreed by both bodies was established almost 20 years later under the auspices of RAJAR.

[57] BBC Press Release, 24 September 1974. BBC WAC Audience Research Commercial Radio R78/3, 861/1).

[58] Director General 1969–1977.

[59] Director General 1982–1987.

[60] Director General 1977–1982, at this time was Managing Director, BBC Network Radio.

arrival of a national presence which was most perturbing, particularly as it had the capability of causing damage to the BBC's other national networks which up until this point had been largely regarded as safe from competitive interference.

Within the BBC the 1990s began with a couple of actions which could be deemed defensive. Radio 2 became the first BBC national service to broadcast exclusively on the FM frequency[61] and a brand new national station began on the AM frequency called Radio 5. Radio 5 was introduced on 27 August 1990 on Radio 2's now redundant medium wave capacity and carried a mixture of sport, children's and educational programmes. Its mission statement appeared nebulous and many considered it a 'dumping ground' of sorts (Crisell 2002, 226), and indeed it was created as a solution to the government's desire to end simulcasting of the same services on both AM and FM. The BBC's Director of Radio at the time, Jenny Abramsky, sums it up succinctly as being a repository for:

> ... the sports output from Radio 2 Medium Wave, all the Schools and Continuing Education programmes from Radio 4 FM, the Open University programmes from Radios 3 and 4 FM and programmes for children and young people from Radio 4 and some World Service output. This was a network with no audience focus, born out of expediency. (Abramsky 2002)

The campaign to relinquish simulcasting across the BBC networks coincided with the 1991 Gulf War[62] which provoked the BBC to introduce a rolling news coverage network on Radio 4's FM frequency with the regular scheduled service continuing on long wave. Although officially named Radio 4 News FM, the service soon became nicknamed 'Scud FM'[63] and its success inspired the BBC to aim to provide a new permanent, national, rolling news network. However, with Radio 4 listeners marching on Broadcasting House to protest against any such network being housed permanently on either of Radio 4's AM or FM frequencies (Hendy 2007, 344) and with the audience for Radio 5 becoming fragmented (Starkey 2006, 24), it was clear where the new service would sit. Thus Radio 5 was replaced by a rolling news and sport network on 28 March 1994 and renamed Radio 5 Live.

---

[61] That is, with no concomitant service on AM.
[62] From 16 January to 2 March 1991.
[63] From the Scud missiles used by the Iraqi forces in the war,

The new found self-confidence of the commercial sector at both a local and national level meant it was now competing with the BBC on an equal footing. But as far as local radio was concerned, the BBC remained ebullient and it was helped by the fact that many of its stations were to be found on the FM band. Not only that but the BBC had been quietly content with its local radio performance vis-à-vis ILR stations. By 1988 the BBC was able to boast that its local radio service was among the most listened to of all local radio services available, with 16 of the 32 English stations reaching at least 30% of their target population and a regular total weekly audience of 10 million.[64] Even the former Director of Radio, Frank Gillard, was able to comment that BBC local radio had 'won its spurs'[65] in terms of its public service responsibilities.

The major area of contention for the BBC as the commercial sector found its feet, at first a local and then a national level, was the impact on its national networks and in particular its most popular station—Radio 1. As Radio 1 entered its 25th year in 1992, its audience figures began to plummet. While the ILRs, the pirates and Atlantic 252, as well as the imminent arrival of Virgin 1215 may be seen as contributing factors, much of the fault lay within the BBC itself as many cultural commentators began to accuse Radio 1 of abandoning its core role of providing a clearly defined music service delivered in an attractive manner. Much of the criticism centred on the DJs, many of whom it was believed had been 'sat in front of the microphone too long.'[66] Despite this, the function of Radio 1 as a provider of both pop and specialist music programming still held 'an exalted position among its audience group' (Morrison 1992) which accounted for 28% of the population, 83% of which were in the 16–44 age group (Morrison ibid). The situation at Radio 1 had developed into a major crisis by 1993, arguably one of the corporation's biggest in terms of faltering networks. It created a tremendous amount of tension internally as leaders began public arguments about how the situation should be properly addressed, with the new Managing Director of Network Radio, Liz Forgan,[67] pledging her support for Radio 1 'continuing in its present form'[68] while Janet Street Porter, BBC TV Head of Youth Programmes,

---

[64] *Ariel*, 27 July 1988.
[65] *Radio Times*, 7 November 1987.
[66] How Radio 1 Has Failed Pop. Tony Parsons. *Daily Telegraph*, 22 September 1992.
[67] Managing Director, Network Radio, 1993–1996.
[68] *The Guardian*, 18 February 1993.

called for a 'mixed network with a greater volume of speech content.'[69] Radio 1 exemplified the:

> ... relationship between the media industries and public 'taste', and...
> the role of popular culture within public service broadcasting
> organisations typified by the BBC. (Hendy 2000)

The atmosphere of despondency surrounding Radio 1's struggle to maintain its identity and its audience was compounded by constant media reports that the government was considering proposals that Radio 1 and Radio 2 should carry advertising[70] or even be sold off completely, as argued vehemently by the Radio Authority in its response to the 1992 Green Paper on the future of the BBC.[71] Relations between the BBC and the commercial sector had reached an all-time low with the latter pursuing a relentless campaign against Radio 1 in the hope that the government might see its privatisation as the solution. The Radio Authority's position on this changed slightly however towards the end of 1993 as a report it commissioned from the Henley Centre hinted at the fact that although there might be benefits to be had from a privatised Radio 1, there was a chance that such a step might have a 'detrimental effect on the sector as a whole due to the possible losses that could be inflicted on ILR and INR should a company within the commercial sector take it over.'[72]

In addition to constant calls for the selling off of Radio 1, the AIRC constantly monitored Radio 1 so as to point out any inconsistencies in its terms under the BBC charter. So, for example, in May 1993 the AIRC was able to accuse Radio 1 of breaching its own licence by bringing in a form of back door advertising to the network, claiming the record industry and other companies were spending £1 million a year on promotions with Radio 1 in return for publicity for their products.[73] The culmination of the huge amount of attention being paid to Radio 1 at the time ultimately led to the resignation of its Controller, Johnny Beerling,[74] in June 1993.

[69] *Daily Telegraph*, 26 February 1993.
[70] *Independent on Sunday*, 27 June 1993.
[71] Green Paper: The Future of the BBC. 1992. Cmnd 2098.
[72] Privatisation of Radio 1. A Report for the Radio Authority and AIRC, Henley Centre, November 1993.
[73] Radio 1 'Backdoor' Adverts Infuriate Commercial Rivals. *Sunday Times,* 16 May 1993.
[74] Controller, Radio 1, 1985–1993. For more on Beerling's time at Radio 1 see Beerling (2015).

If the previous few years had been turbulent, then the next few were to be a veritable maelstrom. Matthew Bannister[75] took over as Radio 1 Controller in July 1993 and set about changing the station's character. Essentially this involved getting rid of long-serving DJs and bringing in more populist and younger presenters as well as specialist music presenters to reflect the various emerging music genres, thus making it distinctive from commercial stations playing popular music. Street (2002, 131) rightly describes Bannister's actions as redefining the station's 'brand image' and branding became a keyword in BBC policy at this time although it was the axing of some of the biggest names in British radio[76] that had the effect of giving enormous press coverage to the medium of radio with headlines such as 'Minor Quake at Radio 1,'[77] 'Radio Ga-Ga'[78] and 'Golden Oldies in Radio 1 "Bloodbath".'[79]

Bannister's initial moves actually proved disastrous as only a year after implementing change the Radio 1 audience hit an all-time low, losing a staggering 3.4 million listeners over the course of the year, and worryingly this drop was almost mirrored by the number of people now tuning into Virgin Radio which itself had overtaken Classic FM as the largest INR station.[80] The threat of Virgin compelled the BBC to produce a special report in May 1994 analysing its competitor's position.[81] The report highlights how Virgin had 'changed its music policy since first coming to air by playing more familiar music and constantly implementing schedule changes in order to achieve a steady growth in audience share' but also concludes that Virgin 'remained less of a threat than the ILR stations and would only pose a serious challenge were it to be awarded a national FM frequency.'[82]

Radio 1's fortunes did turn however once Bannister's changes had settled and he had managed to secure the right presenters for the network, particularly when Chris Evans took over the breakfast show on 24 April 1995. The freefall at Radio 1 came to an end a few months after Evans

---

[75] Controller, Radio 1, 1993–1998.

[76] Particularly names such as Simon Bates, Dave Lee Travis and Bob Harris.

[77] *Evening Standard*, 10 November 1993.

[78] *Today*, 27 April 1994.

[79] *Sunday Times*, 26 September 1993.

[80] RAJAR, Q2 1994.

[81] Virgin 1215: How Much of a Threat is it to Radio 1. Special Report SP93/135, May 1994. BBC WAC: R9/1, 474/1.

[82] Virgin was awarded a London FM frequency on 105.8 in April 1995.

started, as RAJAR figures showed nearly 600,000[83] new listeners tuning in, finally bucking a downward trend and heralding a new era of increasing audiences. However, although this improvement was to continue, the BBC was keenly aware of how the enormous threat from the commercial sector had almost brought Radio 1 to its knees and it was now paranoid about possible future developments within the commercial sector including 'a further expansion of cross media ownership rules, further ILR syndication and the expansion of Virgin.'[84]

While the early 1990s represented a perilous time for Radio 1, the other national radio networks were under pressure too. In order to get a head start against the forthcoming commercial onslaught, the BBC released redesigned logos and branding for all its national networks in April 1990 in an attempt to allow it to 'face up to the challenge within the industry.'[85] Although Radio 2 accounted for 14.7% of all radio listening, its audience had still slipped after losing its medium wave presence in 1991 and it faced a further attack from Classic FM after 1992. At the same time it was subject to calls for it to be privatised too including from a former Director General, Alasdair Milne, who described it as 'the weakest link in the chain of BBC radio networks.'[86] Audience fears that any changes at Radio 2 might promote even more pop music at the expense of specialist music programming had to be allayed by the Controller Frances Line who reassured concerned listeners that these programmes provided:

> ... an irreplaceable service for listeners to and practitioners of many forms of music making which are part of this country's rich cultural heritage.[87]

Radio 3, concerned about the arrival of Classic FM, announced sweeping changes in June 1992 aimed at widening its audience with commentators describing them as the network's 'most radical overhaul since it changed its name from the Third Programme in 1967.'[88] A new Controller, Nicholas Kenyon, was attempting to make Radio 3 more accessible ahead of the launch of Classic FM but yet again, like Frances Line at Radio 2, his main obstacle was persuading the existing audience that this in no way

---

[83] RAJAR, Q2 1995.

[84] UK Commercial Radio. Special Report 95/565C, July 1995. BBC WAC: R9/1, 809/1.

[85] David Hatch, Managing Director, Network Radio. *Ariel*, 3 April 1990.

[86] *The Guardian*, 27 January 1992.

[87] *Daily Telegraph*, 2 May 1992.

[88] *The Guardian*, 30 June 1992.

represented a 'descent into populism' (Carpenter 1996, 341) and that the values they held dear regarding their radio station remained intact.

Radio 4 also went through a period of upheaval in the early 1990s. The issue of siting a proposed 24-hour news network on its long wave band in 1992 caused huge audience intervention and one to which the BBC eventually relented. This proved the Radio 4 audience felt it had a degree of ownership of its station so it was no surprise that a number of minor schedule changes would provoke a similar outcry. These included the moving of *Woman's Hour* from an afternoon slot to a morning slot in September 1991 and the failure of *Anderson Country* in 1994. The uproar surrounding these relatively minor changes caught the BBC off guard which Hendy (2007, 299) explains as being due to the fact that Radio 4 was simply acting as a surrogate for all sorts of wider, deeper disgruntlements regarding questions of permissiveness and political bias:

> In the 1980s and 1990s, these two issues retained their power to offend, and were joined by other, newer concerns—over mediocrity, blandness, superficiality, folksiness, rudeness, aggressiveness, political correctness. Some of these would later be grouped together and labelled 'dumbing down.'

Whether the commercial sector had any role in this process of 'dumbing down,' Hendy does not speculate but it may be reasonable to suggest that during the early 1990s a greater degree of permissiveness had permeated the entire radio spectrum as well as television, thus shifting both media out of the private, domestic realm[89] as had been its considered place and instead into the public sphere[90] (Kay and Mendes 2015, 127).

A significant development in this period was the appointment of John Birt as Director General in 1992.[91] Birt had already been acting as Deputy Director General since 1987 and his first impressions were that there were many problems at the core of the organisation:

> The centre of the BBC seemed stuck in the 1950s… It was quickly apparent that this bloated, bureaucratic monolith was wasting licence-payers' funds on a massive scale. (Birt 2002, 248)

---

[89] Typified as addressing a multi-generational audience in the context of the home.
[90] Described as being consumed by all citizens in the public realm.
[91] BBC Director General 1992–2000.

Birt's response on becoming Director General was to introduce measures to promote economic efficiency across the organisation. Coming from the commercial television sector,[92] he was struck by what he saw as major inefficiencies which he was determined to address, but he was also under pressure from the Conservative government of the day as the BBC's charter was up for renewal in 1996. Birt's response was the introduction of Producer Choice in April 1993—an internal market system requiring BBC producers to buy services from in-house departments or outside suppliers with the aim of promoting efficiency and cutting waste, thus bringing market principles into the internal running of the BBC.[93] Birt was often criticised for using management consultants to formulate his policy with critics sceptical about their understanding of how a large creative organisation like the BBC works, but this overlooks the fact that the very same consultants, McKinsey, were also employed a quarter of a century earlier to achieve improved resource allocation in the reforms of 1967–1972 (Wegg-Prosser 2001). Stoller (2010, 244) claims that at this time while massive change was under way in the BBC, 'radio was treated as being desperately old fashioned.' In fact the opposite seems to be the case according to Birt who described the state of the radio networks at the time as 'in altogether better shape than BBC television' (Birt 2002, 380).

For the BBC the years of the early 1990s were characterised by the fight to keep its national radio portfolio out of danger which meant a lot of change in terms of personalities and content. In many ways the BBC created an internal battleground. Rather than tackling its commercial rivals it seemed to be creating in-house disputes whether with its very core audience over change of programming, or among its own management over change of philosophy, or indeed with its talent as was seen when Classic FM managed to steal five of the biggest names from the Radio 4 programme *Gardeners' Question Time*.[94] Talent also became an issue between BBC stations themselves as different networks competed for various presenters. This was most apparent when Danny Baker walked out of his job at Radio 5 to accept a position at Radio 1 much to the fury of Radio 5 management. The creation of an internal market had internalised competition so that the BBC was competing with external forces while at the same time competing with itself.

[92] Previously Director of Programmes at London Weekend Television (LWT).
[93] For more detailed analyses of Producer Choice, see McDonald (1995), Felix (2000) and Wegg-Prosser (2001).
[94] Gardeners' Defection Time, *The Times*, 10 February 1994.

CONCLUSION

It would appear that this period in British radio history was one that worked very much in favour of the commercial sector. It marks a culmination of the process that started at Heathrow which, with the assistance of legislation in the form of the 1990 Broadcasting Act, the expansion of ILR and the successful introduction of INR, meant that the commercial sector had rid itself of the constraints of independent radio and finally become truly commercial.

While the Reithian approach to Public Service Broadcasting had been weakened by the commercial sector and the Conservative government of the day, it evidently remained a central tenet for the BBC as part of its role to promote a public service ethos. Within the BBC however, this philosophy, while remaining intact, had to work in tandem with another eponymous philosophy, namely, Birtism. The BBC now had to apply market principles to its everyday working in order to achieve the efficiencies required for it to compete with an expanding and ever more resilient commercial sector.

The relationship between the BBC and the commercial sector remained strained during this period. There was little in the way of cooperation over issues beneficial to both parties, and one can argue that the issues surrounding Radio 1 and the commercial sector's intensified desire to compete with it on a local and national level represented a nadir in relations. Many of the disputes between both groups had often been based on the citing of conflicting audience figures as each used separate measurement systems so there was in reality no easy method of comparison or indeed authentication. This was resolved however through a major act of cooperation in 1992 with the establishment of a body called Radio Joint Audience Research (RAJAR) to operate a single audience measurement system for the whole British radio industry.[95] The reason both systems had been difficult to compare was because of the existence of two very different research methodologies with the BBC requiring data for programme evaluation and the commercial sector using its data primarily as a trading currency with advertisers and other investors (Robinson 2000). Although the two methodologies often produced similar results, it was the intensity of competition between the two that necessitated a single

[95] Before this the BBC did its own research, while the commercial sector employed the Joint Industry Committee for Radio Audience Research (JICRAR).

joint service (Robinson ibid). Of course the model is not perfect and since the formation of RAJAR the industry has on numerous occasions been critical of measurement techniques, fearing that if not robust, they may damage radio's credibility among advertisers, regulators and audiences (Starkey 2002). The establishment of RAJAR does however represent an important joint enterprise, albeit one dedicated to furthering competitive rivalry.

Street (2002, 130) says of this period that the 'very sound of commercial radio changed.' Gone was the meaningful speech of the 1970s and 1980s. Instead there was a proliferation of automated playout systems, groups of stations playing the same output leading to what he calls a 'unity of sound' across the commercial network. Street (ibid) also describes a 'unity of branding' which forced some radio critics to lament a 'sameness' of output which turned radio into a branded product rather than a service, but branding was also becoming a cornerstone of BBC Radio as much as the commercial sector. It could be argued that the sound of BBC Radio also changed, particularly that of Radio 1, and it did so in a way to emulate the very sound of commercial radio (Hendy 2000). The question is, did the audience really mind or was this exactly what they wanted? It would seem it was exactly what they wanted as between 1992 and 1995 revenue from advertising in the commercial radio sector almost doubled from £141 million to £270 million, and more importantly, by 1995 commercial radio's audience share overtook that of the BBC for the first time ever (Street 2002, 131)—a landmark event in UK radio history.

So far, we have witnessed the historical divergence which existed in the industry between the BBC and its competitors and which remained largely untouched even after the arrival of legalised competition. We have also seen how, by its very nature, competition between the BBC and commercial radio had created an industry of two halves, both of whom had played their own separate roles in ensuring radio's persistence as a media form. The next major change in the radio world would not come in the shape of new competitive forces but would come in the area of technology. As the 1990s got under way, technical advances had come through the experimental process and were preparing to enter the mainstream, advances which would herald new opportunities for radio. How would these two competing forces react to a new radio technology?

REFERENCES

Abramsky, J. *Sound Matters—Five Live—the War of Broadcasting House—a Morality Story*. Lecture at Exeter College, Oxford University. 31 January 2002. BBC Press Office. (Second in a series of four lectures).

Beerling, J. *Inside Radio 1*. Horning: Lambs Meadow Publications, 2015.

Birt, J. *The Harder Path*. London: Time Warner, 2002.

Briggs, A. *The History of Broadcasting in the United Kingdom. Volume 5: Competition*. Oxford: Oxford University Press, 1995.

Boyd, D. "Pirate Radio in Britain: A Programming Alternative." *Journal of Communication* 36, no. 2 (1986).

Brown, M. *A Licence to be Different: The Story of Channel 4*. London: BFI Publishing, 2007.

Carpenter, H. *The Envy of the World: Fifty Years of the BBC Third Programme and Radio 3*. London: Weidenfeld and Nicholson, 1996.

Collin, M. *Altered State: The Story of Ecstasy Culture and Acid House*. London: Serpent's Tale, 2009.

Cowling, J. "From Princes to Paupers: The Future for Advertising Funded Public Service Television Broadcasting." In *From Public Service Broadcasting to Public Service Communications*. Cowling J. and Tambini, D. London: IPPR, 2004.

Curran, C. (BBC Director General), *The Fourth Television Network*, Lecture at the Royal Institution, 1 February 1974. London: BBC Publications.

Curran, C. *A Seamless Robe: Broadcasting Philosophy and Practice*. London: Collins, 1979.

Crisell, A. *Understanding Radio*. London: Routledge, 1994.

Crisell, A. *An Introductory History of British Broadcasting*. 2nd edition. London: Routledge, 2002.

Edwards, D. (General Manager, Local Radio Development). *Local Radio*. BBC Lunch-Time Lectures, Sixth Series, No.4, 1968.

Felix, E. "Creating Radical Change: Producer Choice at the BBC." *Journal of Change Management* 1, no.1 (2000).

Fletcher, W. *Powers of Persuasion: The Inside Story of British Advertising*. Oxford: Oxford University Press, 2008.

Garner, K. "New Gold Dawn: the Traditional English Breakfast Show in 1989." *Popular Music* 9, no. 2 (1990).

Goddard, G. *Kiss FM: From Radical Radio to Big Business*. London: Radio Books, 2011.

Gordon, J. *Notions of Community: A Collection of Community Media Debates and Dilemmas*. Bern: Peter Lang, 2009.

Gunter, B. "The UK: Measured Expansion on a Variety of Fronts." In *Audience Responses to Media Diversification*. Becker, L. and Schoenbach, K. London: Routledge, 1989.

Hendy, D. "A Political Economy of Radio in the Digital Age." *Journal of Radio and Audio Media* 7, no. 1 (2000).

Hendy, D. *Life on Air: A History of Radio 4.* Oxford: Oxford University Press, 2007.

Hesmondhalgh, D. "The British Dance Music Industry: A Case Study of Independent Cultural Production." *British Journal of Sociology* 49, no. 2, (1998).

Hilmes, M. "Rethinking Radio." In *Radio Reader: Essays in the Cultural History of Radio*, edited by Hilmes, M. and Loviglio, J. London: Routledge, 2002.

Hobson, D. *Channel 4: The Early Years and the Jeremy Isaacs Legacy.* London: I.B.Tauris, 2008.

Horgan, J. *Irish Media: A Critical History Since 1922.* London: Routledge, 2001.

Kay, J. and Mendes, K. "Home Comforts? Gender, Media and Family." In Conby, M. and Steel, J. (eds), *The Routledge Companion to British Media History.* Abingdon, Routledge, 2015.

Johnson, C. and Turnock, R. *Independent Television Over Fifty Years.* Maidenhead: Open University Press, 2005.

Lee, L. "Contemporary Hit Radio." In *Encyclopedia of Radio.* Sterling, C. London: Routledge, 2004.

Lewis, J. *The Audience for Community Radio.* Report for Greater London Council, 1985.

McCain, T. and Lowe, G. "Localism in Western European Radio Broadcasting: Untangling the Wireless." *Journal of Communication* 40, no.1 (1990).

McDonald, O. "Producer Choice in the BBC." *Public Money and Management* 15, no.1 (1995).

Malm, K. and Wallis, R. *Media Policy and Music Activity.* Abingdon: Routledge, 1992.

Mason, M. *The Pirate's Dilemma: How Youth Culture is Reinventing Capitalism.* New York: Simon and Schuster, 2008.

Milne, A. *DG: The Memoirs of a British Broadcaster.* London: Hodder and Stoughton, 1988.

Morrison, D. *The Role of Radio 1 in People's Lives.* Independent report, Institute of Communication Studies, University of Leeds, 1992.

Peters, K. "Sinking the 'Pirates': Exploring British Strategies of Governance in the North Sea, 1964–1991." *Area* 43, no. 3 (2011).

Powell, R. *The Possibilities for Local Radio.* Centre for Contemporary Cultural Studies, Birmingham University, 1965.

Radcliffe, M. *Reelin' in the Years: The Soundtrack of a Northern Life.* London: Simon and Schuster, 2011.

Robinson, L. "Radio Research in Transition." *International Journal of Market Research* 42, no.4 (2000).

Scifo, S. *The Origins and Development of Community Radio in Britain Under New Labour (1997–2007)*. PhD Thesis, University of Westminster, 2011.

Skues, K. and Kindred, D. *Pirate Radio: An Illustrated History*. Stroud: Amberley Publishing, 2014.

Starkey, G. "Radio Audience Research: Challenging the Gold Standard." *Cultural Trends* 12, no.45 (2002).

Starkey, G. "BBC Radio 5 Live: Extending Choice Through 'Radio Bloke'?" In *More Than a Music Box: Radio Cultures and Communities in a Multi-Media World*. Crisell, A. (ed). Oxford: Berghahn Books, 2006.

Stoller, T. *Sounds of Your Life: The History of Independent Radio in the UK*. New Barnet: John Libbey, 2010.

Street, S. *A Concise History of British Radio 1922–2002*. Tiverton: Kelly Publications, 2002.

Taylor, S. ""I Am What I Play": The Radio DJ as a Cultural Arbiter and Negotiator." In *Cultural Work: Understanding the Cultural Industries*, edited by Beck, A. London: Routledge, 2003.

Trethowan, I. (Managing Director, BBC Radio). *The Development of Radio*. BBC Lunch-time Lectures, Ninth Series, No. 4, 14 January 1975.

Wieten, J. and Pantti, M. "Obsessed with the Audience: Breakfast Television Revisited." *Media, Culture and Society* 27, no. 1 (2005).

Wegg-Prosser, V. "Thirty Years of Managerial Change at the BBC." *Public Money and Management* 21, no. 1 (2001).

Worcester, R. and Downham, J. *Consumer Market Research Handbook*. 2nd Edition. New York: Van Nostrand Reinhold Co, 1978

CHAPTER 7

Going Digital: New Technology, New Relationship

The introduction of radio broadcasting in the 1920s employed amplitude modulation (AM) for the transmission of audio signals on medium wave (MW) or long wave (LW) bands. AM radio is the simplest form of radio broadcast and describes a carrier signal whose modulation[1] is varied by its amplitude. It is a method still in use today and employed in not just broadcast radio but also in areas such as citizen band radio and aviation due to its ability to be received under weak signal conditions. It has the advantage of being able to be transmitted over great distances although it can also suffer from interference. Due to the use of narrow bandwidth and the unavoidable crowding on the AM band, sound quality can generally be poor and coverage can vary greatly depending on the time of day and season.

Frequency modulation (FM) first appeared in the 1940s and it modulates the frequency of a signal while keeping its amplitude constant. When the frequency is modulated, music or talk is transmitted via the carrier frequency with the effect of improving the fidelity of the radio signal. The distance ranges for FM are much more limited than for AM, but by operating at higher frequencies this provides for a clearer sound and particularly for music. Although FM signals can be subject to interference, a relatively noise-free signal can be maintained through its use of wider channels and even more so when broadcast on very high frequencies (VHF).

---

[1] For a good explanation of the technology behind modulation, see Poole (1998).

© The Author(s) 2018
JP Devlin, *From Analogue to Digital Radio*,
https://doi.org/10.1007/978-3-319-93070-1_7

Both these methods dominated transmission and reception technology up until the 1990s. But after this point AM and FM were joined, although not necessarily superseded, by an alternative modulation scheme, namely, digital. Yet another attempted solution to what Scannell (2010, 11) describes as 'the very old problem that never goes away—maintaining the quality of the listening experience.'

THE DIGITAL IMPERATIVE

Digital radio broadcasting differs from analogue AM and FM by bundling radio channels into multiplexes and by the use of digital encoding to maximise the use of bandwidth available. Limits to the expansion of radio services in their analogue form have been set by a scarcity of space on the electromagnetic spectrum which is easily rectified by digital radio which can squeeze six or seven full-scale services into a space previously occupied by just one. In the digital system an analogue audio signal is digitised, compressed and transmitted using a digital modulation scheme which transfers a digital bit stream over an analogue channel.[2] By encoding a series of ones and zeros[3] on radio waves themselves, a much more powerful signal can be generated, and with such a signal the wave can carry much more information. The effect of this is to bestow on the digital signal a number of significant advantages over analogue and particularly FM. These include a clearer digital sound quality through the reduction of hiss and crackle, a greater listening choice through the potential of delivering a greater number of stations and an addition of data services, meaning text and other data appendages can be carried alongside the audio signal. In addition to these umbrella characteristics, there can be extra benefits for the listener or user experience derived from additional features which include ease of finding stations at the touch of a button and the ability to pause and rewind live radio.[4]

Although the technology of digital radio is readily definable, the term 'digital radio' as used in the context of this chapter requires clarification, given that different descriptions of means of reception can be employed

[2] For a good introduction to digital technology, see Lax 2009, Chapter 5.

[3] This binary system is the basis of all digital broadcasting.

[4] All these attributes are highlighted by both the BBC: http://www.ukdigitalradio.com/advice/thebasics/ and the commercial radio sector: http://www.bbc.co.uk/reception/radio/digitalradio/.

loosely or even interchangeably. Thus, we find a number of platforms including digital television, internet radio, satellite radio and Digital Radio Mondiale (DRM). This text concentrates on another particular platform—DAB radio—which complies with all the attributes of other digital platforms but, unlike the others, continues an important linkage with radio's inherent portability, accessibility and design and therefore represents the closest link to the traditional mode of radio listening which typified the analogue era in the form of the radio set. It is this notion of 'multi-location reception' (Lax et al. 2008) or what Crisell (1994, 11) describes perhaps more aptly as 'flexibility' which sets DAB apart from other platforms and which makes it more akin to the analogue model which has formed the basis of previous chapters.

DAB as a technology was originally developed in 1981 at the Institut für Rundfunktechnik in Munich, a research centre run by a number of German, Austrian and Swiss broadcasters aspiring to achieve a degree of standardisation of broadcasting technology, and it was this enterprise which would go on to form the basis for the Eureka 147 research project. The oddly titled Eureka 147 project has quite an unostentatious meaning; it signifies the 147th project of the Eureka project established by a Conference of Ministers of 17 countries and members of the Commission of the European Community. The 147th project, which began in January 1987, had as its aim the 'development of a European technical standard for digital audio broadcasting'[5]; the unequivocal driving force was to provide a new impetus for the European consumer electronics industry and those spearheading the advance included a number of European broadcasters (including the BBC), as well as some equipment manufacturers, car makers and transmission companies, and this was supported by the European Broadcasting Union (EBU). O'Neill and Shaw (2010, 36) claim:

> The guiding assumptions underpinning the development of Eureka 147 DAB were that a robust and mature technology developed within Europe's highly regarded high technology research environment would provide an ideal replacement standard for the broadcasting industry.

This interpretation may be over-generous with regard to the ambitions for broadcasting. It may be more appropriate to say at this point, in the

---

[5] Eureka Project Form, 17 December 1986.

late 1980s, that the bedrock of the Eureka 147 project was a desire to boost the competitive edge of the European electronics industry by placing Europe at the vanguard of the digitalisation of radio 'through the development of its own particular broadcast model' (Lax 2003). Such a model would aim to:

> Stimulate a virtually saturated market with new products for car and domestic audio broadcasting units … this will encourage considerable innovation from European microelectronics manufacturers … will provide a long term counterbalance to the increasing dominance of the countries of the far east in the consumer investment goods industries.[6]

Stoller (2010, 149) believes that the involvement of broadcasters stemmed from a different necessity when he claims the main motivation for the BBC, as the sole UK member of the consortium, was 'the hunger among British broadcasters for more spectrum in an increasingly crowded market place.' The arrival of DAB did occur at a time when the UK radio industry was going through enormous change in terms of its structure and position within the media environment. The end of the BBC monopoly at a local level had been well established since the 1970s, and a challenge to its hegemony at a national level coincided with the emergence of the DAB standard. The continuing growth of commercial radio in the UK brought with it an increasingly limited availability of the FM spectrum, which meant that the commercial sector could only anticipate future incremental growth of analogue radio by curtailing its expansive aspirations. So, with the BBC having a leading role and influence in the Eureka 147 project, the future for DAB seemed propitious. In fact, when it was first introduced, there was a general feeling among most broadcasters and the EBU that it 'would replace FM in a short and smooth transition' (Ala-Fossi et al. 2008).

Steps towards DAB in the UK continued with the setting up of the UK DAB Forum in 1993, announced by the Department of Trade and Industry (DTI)[7] which confirmed that DAB was now the preferred technical choice of the government for the future of radio broadcasting. The UK DAB Forum was largely dominated by technical and administrative personnel from the DTI and the BBC and its remit was the promotion of DAB among consumers, not just consumers in terms of consumers of

---

[6] Ibid.

[7] National Forum to be established to Promote Digital Audio Broadcasting. Department of Trade and Industry Press Release, 16 February 1993.

audio—that is, listeners—but also consumers in the sense of potential purchasers of the new required receiving sets, without whom the project would be doomed to long-term failure. It was only with the formation of the UK DAB Forum that the AIRC was:

> ... invited to join at a slightly later stage, certainly after the BBC ...
> whereas national state/public service broadcasters were involved
> in the initial policy discussions, only when the essentials of the
> system had been agreed were commercial and local broadcasters
> invited to participate. (Rudin 2006)

In 1995 the government published a White Paper 'Digital Terrestrial Broadcasting: The Government's Proposals'[8] which eventually would lead to the 1996 Broadcasting Act.[9] In the debate on the bill as it passed through Parliament, Virginia Bottomley, the Secretary of State for National Heritage, told the House of Lords:

> We stand on the verge of a new broadcasting revolution even
> more significant than the change from black and white to colour
> television.[10]

The Act itself outlined provisions for licences and the operation of multiplexes and provided a very attractive incentive for commercial stations to embrace digital. It allowed for one national digital multiplex to be operated by BBC and one by the commercial sector as well as 26 local and regional multiplexes. The INR stations would be guaranteed a presence on a national multiplex and more importantly, any ILR station which agreed to broadcast on a local multiplex would get an automatic eight-year extension to its analogue licence.[11] This was an enormously generous inducement for the commercial broadcasters to step on to the digital bandwagon, but Stoller (2010, 280) is very critical of the government's actions in this regard:

> The idea of effectively incentivising the radio companies to join DAB ...
> effectively locked in the major radio companies to apply for DAB licences ...

---

[8] Digital Terrestrial Broadcasting, The Government's Proposals, 19 September 1995. Cm: 2946.
[9] Broadcasting Act, 1996. London, HMSO.
[10] Hansard: *House of Lords 16 April 1996. Vol 275 cc537-614.*
[11] Broadcasting Act, 1996. London, HMSO.

it also forced the entire industry to become supporters of DAB, when some commercial scepticism would have been useful.

The Act made little provision regarding content and left regulation of this aspect to the owners of the multiplex licences who were able to simply simulcast existing services, thus giving real power to the established main commercial players at the detriment of smaller companies. This provided further incentive for the commercial sector now tantalised by the double-headed enticement of automatic extension of existing analogue licences and the absence of strict content regulation. Rudin (2006) believes these factors:

> ... amounted to very generous concessions and resulted in further
> consolidation of an already rapidly consolidating commercial
> radio system.

The 1996 Broadcasting Act provided impetus for greater fortification of the position of the commercial sector and brought it firmly into the DAB arena, largely because the commercial companies could envisage the benefits, as Lax (2007, 109) concludes:

> It would appear the UK government in its desire to secure the
> the growth of digital radio needed to encourage the commercial
> sector to participate and offering inducement allowed that sector
> to expand and strengthen its position particularly against the BBC.

The effect of this new-found boost in self-confidence from within the commercial sector was to spur the BBC into even greater action and this too resulted from the provisions of the 1996 Broadcasting Act. Although having run a DAB experimental service since 1993, the Act now granted the BBC spectrum to operate a national multiplex to simulcast its existing national networks,[12] and, as Starkey (2008) observes, the BBC probably felt 'it could not afford to be left behind at this crucial point in digital radio history in the UK.' It would be both the BBC and the commercial sector who would play unique and joint roles in implanting the very notion of DAB in Britain over the course of the 1990s.

---

[12] Broadcasting Act, 1996. London, HMSO.

THE EARLY 1990s: CONSOLIDATION OF DAB

In July 1991 over 500 people from the radio industry, the consumer electronics industry and the press attended the very first UK demonstration of DAB by BBC engineers in Birmingham. The demonstration included a 20-minute coach journey around the city to illustrate the 'ruggedness of the DAB system in a city centre environment'[13] and the BBC concluded it was received with great acclaim, with comments ranging from 'stunning' to 'mind-blowing.'[14] Following this positive reaction, the BBC carried out its first major market analysis of the potential of DAB during the course of 1992.[15] The subsequent report highlights the existing, well-trodden attributes which formed the core of the driving imperative to promote DAB, namely, its innovative nature and its potential for providing more frequency space, but the report is also realistic about DAB's inherent deficiencies. Reliability of reception and sound quality were still not proven selling factors, with more than 90% of listeners voicing satisfaction with the existing FM reception. In addition to this, the prohibitive cost of first generation sets would further act as a disincentive.[16] The BBC also reveals at this stage its perspective on how the commercial sector will likely react to DAB, stating that 'the cost implications suggest that it is therefore in the best interests of ILR/INR to prevent or slow the introduction of DAB regardless of any benefits it may bring consumers.'[17] At this point it is evident that the BBC does not envisage any early participative role for the commercial sector in promoting DAB and nor does it see any benefit in such a role as 'even with involvement of the BBC, INR and ILR, take up will be slow.'[18]

The BBC ran its first short-term DAB trial in London on 6 September 1993 and this subsequently set in motion actions by the BBC Board of Management Technical Committee on 10 December 1993 to trial a longer-term DAB service in the London area beginning in 1995 which

[13] ENGINF: The Quarterly for BBC Engineering, Technical and Operational Staff. No. 45, Summer 1991

[14] Ibid.

[15] DAB: A Market Analysis. BBC Policy & Planning Unit. 5 October 1992. BBC WAC: E120-014, Part 1, Digital Audio Broadcasting—Project Group.

[16] The report suggests people will still be buying non-DAB radios well beyond 2000 (ibid.).

[17] Ibid.

[18] Ibid.

would deliver the existing national radio networks, and it was hoped this would then lead to an extension of the service to 60% of the UK population by March 1999.[19] It was hoped this initiative by the BBC would encourage manufacturers to join the cascade and ultimately supply sets for the market. The BBC's commitment at this point cannot be underestimated as it was to invest an initial £8 million to launch the project with an estimated subsequent operating cost of £1 million per annum thereafter.[20] It was apparent that the BBC would be the early driver of DAB in the UK and that it would do so on its own without the support of the commercial sector. Correspondence between the BBC and the government demonstrates this point:

> We have embraced and developed the new technology because we believe it offers very significant long term benefits for our listeners and for our radio broadcasters in general … There are risks in being *the leader* for such a venture … but the early introduction of DAB will be of immense value to our listeners and also, offers the UK industry as a whole the valuable opportunity to become pioneers in the development of DAB products in Europe.[21]

At this point it is reasonable to question why the BBC did not simply continue with the roll-out and strengthening of FM since, after all, a policy of conversion of the popular national networks had been taking place around the same time. Radio 1 began occasional national broadcasts on FM in 1988 by being allowed to use Radio 2's FM transmitters for a few hours each week.[22] Once the 97–99 MHz frequencies became available towards the end of the 1980s, Radio 1 acquired them for its own national FM network. It was estimated the likely loss of audience, resulting from the complete switch from AM to FM on Radio 1, would be in the region of between 120,000 and 972,000 since BBC Engineering believed that although 98% of the country had FM coverage only 96% of the UK

---

[19] Note from Bob Phillis (BBC Deputy Director General) regarding Board of Management Technical Committee meeting, 10 December 1993. BBC WAC: G080-006 Digital Audio Broadcasting.

[20] Ibid.

[21] Letter from Patricia Hodgson (Director, Policy & Planning) to Paul Wright (Department of National Heritage), 18 February 1994. BBC WAC: G080-006 Digital Audio Broadcasting.

[22] 10:00pm to midnight on week nights, Saturday afternoons, as well as Sunday evenings, most notably for the Top 40 Singles Chart countdown.

population would receive it satisfactorily. In the end the actual loss was calculated to be more in the region of at least 500,000; thus the switch to FM had a harmful effect on the Radio 1 audience.[23] Radio 2 had similarly lost a significant number of listeners (2 million) when it surrendered its AM wavelength to Radio 5 in August 1990.[24] Although a relatively robust technology and with a successful market penetration, FM still faced the major obstacle of 'spectrum scarcity.' The FM spectrum was a finite resource and required 'rigorous protection against interference from competing broadcasters through government licensing of the spectrum for exclusive usage' (Berlemann and Mangold 2009, 2). As competition in the radio industry escalated with the arrival of the first national commercial broadcasters between 1992 and 1995, there were fears within the BBC that any new FM frequencies that might become available would be allocated to an expanding commercial sector or, worse still, existing BBC frequencies would be simply handed over to new commercial services. This, combined with a recent spate of success within the commercial sector, made Managing Director of BBC Radio, Liz Forgan, come to the conclusion that digital radio itself might even take further audience numbers from the BBC to the commercial stations, so it was essential to attempt an early digital impact.

It was still felt that the commercial operators were likely to eschew any steps to aid in the early implementation of DAB. Despite misgivings regarding any short-term commitment from the commercial sector, an interim report[25] nevertheless reveals the BBC's decision to initiate ongoing discussions with the Radio Authority and INR stakeholders to develop UK DAB services with the aim of benefitting the entire radio industry. The report also outlines the BBC's less than altruistic reasons for attempting to coax the commercial sector. RAJAR research showed the number of radio stations available to the average listener had grown from around 5 to 14 over the previous 20 years and over the same period the BBC had been steadily losing its listening share to the commercial sector. As this trend was likely to continue, one of the BBC's biggest challenges would be to keep listeners loyal to the corporation, particularly as the

[23] BBC Radio 1FM: The Switch to FM Only Broadcasting. Report by Paul Robinson (Managing Editor, BBC Radio 1FM) 1994. BBC WAC: E120-014 Part 1 Digital Audio Broadcasting—Project Group.

[24] Ibid.

[25] Digital Audio Broadcasting Report, 29 September 1994. BBC WAC: G080-006 Digital Audio Broadcasting.

commercial sector was likely to expand at a greater rate than the BBC in terms of services offered. With the commercial companies constantly pressing for the BBC to hand over one or more of its high-quality FM frequencies, the simple conclusion was that 'without a DAB commitment and the accompanying future prospect of unlimited spectrum they may eventually succeed.'[26] If DAB was the driving imperative in order to increase potential spectrum, then the commercial sector would surely come on board for 'the final drive to completion.'[27]

When ILR first launched in 1973 it broadcast its services on both FM and AM, following a pattern of simulcasting which the BBC was also pursuing. It can be argued that this simulcasting on two wavebands may have held back the more widespread penetration of FM until the 1980s when both the BBC and the commercial stations were forced by the government[28] to end simulcasting and instead offer different services on their FM and AM wavelengths. Also, in 1990 the Radio Authority began to license more local FM-only stations which sought to cater for diverse audience groups.[29] As well as this, the continued growth of music radio in the UK—largely at the hands of new commercial stations—combined with the gradual falling price of FM receivers (increasingly in small portable transistor radios) helped contribute to the inexorable rise of FM radio.

Commercial radio success into the 1990s on FM led operators to press for the BBC to hand over one of its FM frequencies such as Radio 4's FM position. For some within the commercial sector the struggle with the BBC over FM frequency space was becoming irksome, and DAB was imagined by the newly created Radio Authority in 1990 as representing 'the chance to get even' (Stoller 2010, 275). The BBC had retained the best AM frequencies and dominated the FM bands. DAB would even the score in the mindset of the Radio Authority, which then set about a process of technical planning for the introduction of DAB for commercial radio in the UK in an attempt to overcome what the Radio Authority's Director of Engineering, Mark Thomas, saw as 'the failing of analogue broadcasting for the commercial sector' (Thomas 1995).

---

[26] Ibid.
[27] Ibid.
[28] Green Paper: Radio Choices and Opportunities (Cm 92), 1987.
[29] Examples in London included stations like Kiss FM, Choice FM and Jazz FM.

DAB was to represent more than a simple technology resolution how-ever. It could also play a role in creating a greater diversification of content. By the 1990s, although there were more radio stations than before, most of them were targeting the same audience and so sounded very similar. Tacchi (2000) cites research which supports the notion that broadcasting in the UK by the mid-1990s was suffering from 'a lack of innovation in and increasing blandness of local commercial radio.' In order to continue growth and at the same time broaden the variation of content, commercial radio would require even more stations and by the mid-1990s it was clear the amount of spectrum available on FM would not serve these ambitions.

THE MID-1990S: EXPERIMENTATION AND CIRCUMSPECTION

In the run up to the launch of the BBC's London experimental DAB net-work in 1995, which subsequently became known internally as the London Experiment, the corporation remained realistic yet sanguine about the prospect for the new technology. The months up to the launch would focus on manufacturers rather than consumers:

> 1995 is not the year DAB will take off as far as the public is concerned, they need the availability of radio receivers and since mass marketing of these products is unlikely to take place before 1997, the BBC's marketing role up until then should not be aimed at licence payers. Instead during this early experimental stage, BBC DAB should focus its resources at supporting the Corporation's switch-on in September 1995, and at encouraging and stimulating manufacturer development.[30]

Active steps were being made to persuade receiving set manufacturers to embark upon investing in making DAB sets. The BBC contacted Philips[31] regarding the provision of receivers for test purposes for the forthcoming experiment phase and secured and ordered 12 such receivers which Philips duly agreed to supply by March 1995[32]—these sets may very

[30] BBC DAB Marketing Strategy (1995–1998), 1994. BBC WAC: E120-014, Part 1, Digital Audio Broadcasting—Project Group.

[31] One of the world's largest electronics companies http://www.philips.co.uk/about/company/index.page.

[32] Proposed DAB Service. Note from Rod Lynch (Managing Director, BBC Resources to Liz Forgan (Managing Director, BBC Radio), 28 March 1994. BBC WAC: G001-002 Digital Audio Broadcasting.

well be considered to have been the first bulk order for DAB sets in the UK. The DTI also played a role in attempting to persuade UK manufacturers to help develop the required integrated circuits or 'chips' necessary for the mass production of DAB receivers through the offer of financial support to interested companies,[33] no doubt because this would have been seen as a lucrative pathway for British industry.

When this joint BBC and DTI campaign to engender interest amongst British manufacturers was instigated, it was noted that interested parties were 'thin on the ground' and that the BBC might have to invest further funds in becoming part of some form of consortium with those who could be persuaded to participate.[34] Manufacturers remained reluctant to make any commitment to providing either receivers or components of receivers for the wider market and were clearly awaiting positive signals that DAB might become a reality. Acting as the sole promoter of DAB in the UK at this time, the BBC found itself spearheading the new technology's development. It appreciated the commercial sector would be in no position to risk investment at this stage, and therefore had to devote a significant amount of effort to getting receiver manufacturers on board and this long before it envisaged the wider audience even contemplating the move to digital.

Liz Forgan was able to formally announce at the Voice of the Listener and Viewer luncheon on 24 November 1994 that the BBC was to launch its digital audio broadcasting service in September 1995.[35] In addition to the existing five national BBC radio services, there would initially be extended parliamentary and sports coverage, all offering near-CD quality audio:[36]

> DAB offers the chance for listeners to have access to parliament
> on a scale which the BBC has never before been able to offer ...
> good reception will no longer depend on adjusting aerials or
> moving receivers. The Proms will be crystal clear, even in a car.[37]

---

[33] Ibid.

[34] Digital Audio Broadcasting (DAB) Service Project: An Update on Progress, 1 June 1994. BBC WAC: G001-002 Digital Audio Broadcasting.

[35] Engineering test transmissions had been in operation since 1990.

[36] BBC Radio to Launch New Digital Service in 1995. BBC Press Release, 24 November 1994.

[37] BBC Makes Radio's Future Crystal Clear. *The Times*, 25 November 1994.

Forgan went on to add the BBC was the first broadcaster in the world to make such a firm commitment to launching a DAB radio service, while the Department of National Heritage also confirmed government plans for five commercial stations to follow the move to DAB. However, this air of digital exuberance was tempered by the Dutch electronics company, Philips, which pointed out that in the immediate future radio manufacturers would not be able to meet the demand for DAB receivers and that any mass-produced versions were still likely to cost hundreds of pounds.[38] Despite this, the BBC remained committed to the launch, which would only cover the Greater London area initially but with the intention of reaching 60% of the UK population over the following three years.[39]

There was still a great deal of uncertainty even after the BBC had made the announcement about the London Experiment. The BBC had crossed the Rubicon by committing to a trial which it was hoped would become permanent. However, it knew that beyond this point the future success of DAB would depend on having two groups on board: consumers and eventually the commercial radio sector. The audience had now become known as consumers in much of the BBC documentation of the time as it was their purchasing power which was of greater necessity than their listening tastes. Equally, the commercial companies were required to help push DAB as a distribution and reception standard in order that these consumers would have access to all digital radio services, both BBC and commercial. But this led to the core issue at the time, which was what the BBC called the 'chicken and egg problem'[40] in getting consumers and radio companies to promote DAB. Which group would be first to take action and essentially force the other to come on board? For the commercial companies it would mean incurring significant costs in transmitter upgrades as well as the lesser recognised problem of coordinating a number of other stations which would be required to fill a DAB multiplex. For consumers, having only the BBC national stations on DAB would not be a great enough incentive to purchase a new, expensive receiving set as many would indubitably prefer to wait until their favourite commercial station was also available on the same platform.

---

[38] *The Times* (ibid.).

[39] BBC to Launch Digital Radio Services in 1995. *Broadcast,* 2 December 1994.

[40] Discussion Document: Background & Development of DAB, 3 February 1995. BBC WAC: E120-014 Part 1 Digital Audio Broadcasting—Project Group.

One can argue that the period up until the launch of the London Experiment was characterised by a slight degree of uncertainty within the BBC, with possible scenarios of success or failure undergoing constant evaluation, the worst possible outcome being DAB technology written off completely. Crucial to these potential outcomes was whether the commercial sector would eventually engage wholeheartedly in the process, with 'no interest or investment emanating from that sector being the worst possible outcome.'[41] The inherent dangers of any substantial delays in DAB roll-out was also a contributing factor in the BBC's feeling of insecurity, unless DAB was to generate some success in the medium term it could rapidly spiral into failure. Since the BBC clearly understood this to be too much of a financial risk for the commercial companies at this stage, it concluded that it still had more work to do on its own in order to promote wider uptake of DAB and now identified a possible strategy for continuing the drive to digital, which was to provide entirely new services which would appeal to new audiences who would only be able to receive such new services on new DAB sets. This competitive element might also have a coercive effect on competitors. Having considered these options, the BBC came to the conclusion that in order to achieve a targeted 40% of penetration in 10 years, DAB as a whole must carry distinctive new services.[42] Alas the notion of new services and what form they would take would not come to fruition for another seven years.

Unfortunately, any confidence over the experimental transmissions was crudely deflated at the 1995 Internationale Funkausstellung (IFA)[43] in Berlin at which a number of European manufacturers had intended to display receiving sets in what was supposed to be in effect a launch of the first models. Cornell (2003) explains that it became apparent just a few weeks before the exhibition that the sets would not be forthcoming but the BBC decided to go ahead with the launch of its DAB trial. This was a hugely disappointing setback for the BBC and for DAB radio generally in Britain. Transmissions were to begin but there was no way to receive them. The digital reality for the foreseeable future was that the BBC would be broadcasting on DAB 'to virtually no one outside the technicians within the broadcasting industry' (Cornell ibid.).

---

[41] Digital Audio Broadcasting: Strategy & Business Plan, 14 July 1995. BBC WAC: E120-014 Part 1 Digital Audio Broadcasting—Project Group.

[42] Ibid.

[43] Originally the Berlin Radio Show, the exhibition is now one of the world's leading trade shows for consumer electronics. http://b2b.ifa-berlin.com.

The BBC eventually launched its DAB trial service on 27 September 1995, claiming that the 'BBC embarks on the third age of radio with Digital Audio Broadcasting.'[44] Liz Forgan heralded the trial as:

> ... the dawn of a third age of radio—the technological progression from AM which is now 100 years old, and FM, now 50 years old, into the digital multi-media world of the twenty-first century. Consumers will get superb quality sound, a fade-free signal and a whole range of new services on simple, easy-to-use sets.

The initial service would cover 20% of the population using 5 transmitters in the Greater London area, with the intention of extending to 60% of the population by March 1998 as a further 22 transmitters were planned to become operational.[45] As well as simultaneous digital transmission of the existing BBC national analogue stations and the BBC World Service, extra channels were introduced although these could be considered as bolt-ons rather than new channels. One was a sports commentary service, another was a BBC Parliament channel.[46]

With continued reticence from manufacturers, the involvement of the commercial sector was beginning to become a greater imperative if DAB was to succeed, and so meetings were initiated with the Radio Authority in order to discuss the shared interest both parties might have in promoting DAB. BBC notes from an initial meeting in October 1995 suggest there was mutual support for a greater remit for the UK DAB Forum as well as the instigation of a 'DAB champion' from the commercial side.[47] The first meeting of the Publicity and Promotions Working Group of the UK DAB Forum took place at the BBC on 4 January 1996, and among those present were representatives from the BBC, from set manufacturers, from transmitter companies as well as from the AIRC. The aim was to:

> ... actively promote awareness of DAB in the UK with a coordinated message to specific target audiences particularly retailers and the trade press.[48]

[44] BBC Press Release, 27 September 1995.

[45] Ibid.

[46] This channel continued until November 2000 when it transferred to the digital terrestrial television platform.

[47] Minutes of DAB Project Board Meeting, 27 October 1995. BBC WAC: E120-014 Part 2 Digital Audio Broadcasting—Project Group.

[48] UK DAB Forum—Publicity & Promotions Working Group, minutes of meeting on 4 January 1996. BBC WAC: E120-014 Part 2 Digital Audio Broadcasting—Project Group.

The BBC was desperate for this to work as DAB now formed a core part of BBC Radio's 10 Year Strategy with an implementation cost of £23 million cited in its DAB business plan.[49] The 1995 IFA show represented a low point for the BBC in its DAB journey as it had put tremendous faith in the manufacturers to produce the essential sets for display. From this collapse in communication and faith, it is possible to argue that the BBC then shifted its focus more towards the commercial sector as a more reliable partner to help launch DAB.

Commercial radio's concern over a lack of spectrum on FM found some relief however in ongoing research which pointed to its healthy status in terms of audience figures, with the number of commercial stations increasing to the point where the BBC had steadily been losing listening share to commercial competitors over the previous 20 years.[50] A cursory inspection revealed that DAB offered to resolve the long-standing spectrum problem and at the same time permit the commercial sector to expand its presence, particularly against the BBC, and thus a survey of DAB was set in motion. As early as 1990 the Radio Authority had begun a process of technical planning for the introduction of DAB in the commercial sector (Stoller 2010, 275). The AIRC also established a DAB Committee which first met on 30 August 1991. From this meeting one can ascertain that the companies were keen on the idea of DAB and what potential benefits it might bring, but there was also a note of caution regarding DAB penetration as exemplified by Peter Jackson, the Chief Engineer at Capital Radio:

> Stations should be allowed to simulcast DAB programmes with their AM and FM programmes until they determine sufficient DAB penetration has been achieved to rely on it as the sole transmission medium.[51]

These early steps culminated in the Radio Authority's first policy statement released in October 1992:

> The Radio Authority believes that DAB is the single most significant advance in sound radio transmission technology since the development of FM broadcasting; its impact is likely to be even greater than that of FM.[52]

---

[49] Ibid.
[50] RAJAR 2014.
[51] AIRC, DAB Committee minutes, 30 August 1991.
[52] Initial Policy Statement on Digital Audio Broadcasting. Radio Authority, October 1992.

This represented the commercial sector's first formal stance on DAB and goes some way to suggest an unconditional support for DAB broadcasting and the future place of that technology in the UK radio industry. But how far does this commitment go beyond mere words? The BBC thought it was no more than a statement of intent but without any foundation in the form of action, in fact, in reacting, the BBC believed that this statement did not reflect the more realistic position which was that at this stage the required transmitter investment cost for the commercial stations would be wholly prohibitive.[53] The BBC took its analysis a step further, concluding that ILR and INR stations would fear the fact that DAB would bring new competition rather than additional income and that it was therefore

> ... in the best interests of ILR/INR to prevent or slow the introduction of DAB regardless of any benefits it may bring consumers. If the Radio Authority and/or the BBC do not push for DAB, inaction would probably be the highest reward strategy for ILR/INR ... Although ILR/INR and the AIRC are being very positive about DAB, when the time comes, actions may not match words.[54]

Throughout 1993 and 1994 it was deemed the commercial sector's stance had not changed. The AIRC continued with its support for DAB in public, but BBC research suggested genuine support remained lukewarm particularly amongst ILRs although one particular report paints a more positive picture at INR level by claiming that Virgin Radio had expressed an interest in having a national DAB service and, if the BBC were to decide to launch a national service, they may want to start at a similar time.[55] For the BBC, as the UK's digital pioneer, this was the first encouraging feedback from a major player within the industry and motivated the corporation to continue on-going discussions with both the Radio Authority and individual INR stations in a two-pronged strategy 'to develop UK DAB services.'[56]

[53] DAB: A Market Analysis, BBC Policy & Planning Document, 5 October 1992. BBC WAC E120-014 Part 1 Digital Audio Broadcasting—Project Group.
[54] Ibid.
[55] Digital Audio Broadcasting: A Note by Henry Price, BBC Engineering, 10 December 1993. BBC WAC: Digital Audio Broadcasting G080-006.
[56] Digital Audio Broadcasting Report, 29 September 1994. BBC WAC: G080-006 Digital Audio Broadcasting.

The envisaged changing radio landscape was appreciated by both the BBC and the commercial companies by the mid-1990s, not just for each in relation to the other but in terms of a general picture for the medium as a whole. BBC research predicted that over the course of the decade, general competition for radio listeners' time would increase and this would not just come from within the BBC/commercial sector dynamic, resulting from an increase in the number of stations. Instead, alternative services, particularly in relation to music, would also represent a wider threat— these included cable and satellite television, CD technology and embryonic alternative technologies such as video on demand.[57] Howard (2004) agrees these concerns were also at the forefront of the minds of some within the commercial sector 'who had an eye on the idea of future consolidation within the radio market.' It was thought the proportion of households which would be in a position to receive alternative music services would increase by more than 50% over the following ten years.[58] For the commercial sector it was now becoming important to analyse what DAB could offer in this scenario, and to that end representatives of both the Radio Authority and the AIRC attended a BBC DAB seminar at Kingswood Warren[59] in late 1994, outlining the BBC's own plans and expectations.[60]

Stoller (2010, 277) claims that by this time both the Radio Authority and the BBC had firmly 'nailed their colours to the mast' regarding their stance on DAB, but it is important to recognise that the Radio Authority's posture existed only in policy and not in practice and there was not yet a concerted cross-industry drive towards DAB, in effect the BBC was still leading and the commercial sector merely following. In 1994 the Radio Authority commissioned a study from GEC-Marconi into how the DAB spectrum might be exploited.[61] Stoller (ibid.) claims this report caused friction among ILRs over the costs which would be involved in switching to DAB transmitters in order to create new services. To resolve this dispute the Radio Authority decided the local DAB areas should replicate existing analogue areas, thus avoiding the cost of new transmitters, this would be cost effective and give the ILRs the opportunity to 'dominate

---

[57] Ibid.
[58] Ibid.
[59] Site of BBC Research and Development.
[60] 6 December 1994. BBC WAC: G001-005-002 Digital Audio Broadcasting—General.
[61] Radio Authority, Annual Report, December 1994.

digital radio' at the local level (Stoller ibid.). So the commercial sector was beginning to embrace a more combative approach in seeking out a position of strength.

As time went on, the BBC feared the worst possible reaction of the commercial sector would be that of losing completely the existing interest it had and ultimately declining to become involved in the DAB process,[62] but it would appear that the commercial sector had bought into the potential advantages of pursuing DAB, and as we have already pointed out—but is worth repeating—another salient factor which was critical to the commercial sector's tentative steps towards DAB around 1995 was the undeniable fact that commercial radio was now in the ascendancy. The number of commercial stations had leapt from less than 50 in the mid-1980s to over 150 by the mid-1990s, but the most significant aspect of commercial radio's success and one that would give it the necessary confidence to pursue a DAB strategy was that by 1995 its share of the audience overtook that of the BBC for the first time (Street 2002, 131). The commercial sector, if it was still taking a back seat regarding DAB, was at least by now reaching into the front.

Lax (2014) believes there existed within the commercial sector during these planning years of the mid-1990s a general, altruistic interest in radio developments, including technology. He goes on to make a curious observation:

> Commercial radio was to be included in the early development of digital radio. However, the DAB system did not accommodate commercial radio as readily as its public service counterpart.

It is Lax's idea that it was the DAB system that did not accommodate commercial radio rather than the other way round, which would seem to imply that during the early stage of development of DAB in Britain, the new technology did not match the very raison d'etre of commercial radio. We know that across Europe it was a technology promoted by public service broadcasters since they possessed the required wherewithal to encourage its development (Hendy 2000a, 50), but Lax's observation forces us to re-examine the commercial sector's initial reticent foray into DAB as being due to the very nature of DAB technology itself as opposed to being the result of the commercial sector's own decision-making process.

[62] Digital Audio Broadcasting: Strategy & Business Plan, 14 July 1995. BBC WAC: E120-014 Part 1 Digital Audio Broadcasting—Project Group.

Addressing a commercial radio convention held in Dublin in 1995, Stoller, in his newly appointed position as Chief Executive of the Radio Authority, feared that commercial companies were still too unaware of DAB:

> If it happens, DAB is going to be the major most important single development for radio since the transistor. So while I don't want you to abandon any of your natural commercial caution, you have got to understand DAB and get excited about it. Perhaps above all, you have got to understand how important may be the statutory framework and administrative context for DAB.[63]

What is striking about Stoller's contribution is the continuing seed of doubt reflected in those first few words 'If it happens.' He goes on to explain that the eventual long-term government intention will be to turn off analogue, so it is crucial that commercial stations understand the implications of this and where that will leave them if they don't 'get to grips with the implications of the details of DAB.' Stoller also told those assembled that he had informed the government of the Radio Authority's desire to build on the strengths of the analogue system by simply replicating and extending ILR and INR into the DAB domain.

It is important to note that by late 1995 the commercial sector was taking tentative practical steps to embrace DAB. So in November 1995 the nine local commercial stations in Birmingham became the first UK independent companies to transmit on a live DAB system. NTL launched the transmissions on Thursday 9 November at the Sound Broadcasting Equipment Show in the city with plans to run future experimental services elsewhere in the UK. In historical terms, these Birmingham transmissions are important in that, although short lived, they mark the first significant launch of DAB for commercial radio and reveal a decision within the sector to enter the DAB fray. As 1996 arrived there was a feeling that the commercial sector could and should build on its success, not just in terms of the increasing range of services it could provide but also in terms of its increased audience share vis-à-vis the BBC. Of course the growth of the sector would mean increased competition between companies, but with an expected emphasis on radio as a lucrative advertising medium as well, 'this period of growth should be seized upon.'[64]

[63] Tony Stoller, Chief Executive Radio Authority. Speech to Commercial Radio Convention, Dublin, 1995.

[64] Tony Stoller, Chief Executive, Radio Authority. *The Independent*, 2 January 1996.

There was also a strong message emanating from the government at the beginning of 1996. Heritage Secretary, Virginia Bottomley, wanted ALL broadcasters to embrace new digital technology while appreciating that this would involve a long-term risk investment on the part of British media companies.[65] Bottomley went on to point out that media ownership proposals contained within the forthcoming Broadcasting Act would aid in this process by ensuring continued competition and plurality of ownership which she insisted would help broadcasters to:

> ... serve viewers and listeners, not marauding monopolists. Broadcasting is too powerful and pervasive a medium to give control to any one organisation.[66]

Reading between the lines it is possible to conclude that the government is telling the commercial companies to take a more leading role in the UK media industry, including in the launch of digital platforms, but one can argue that the 1996 Broadcasting Act facilitated this more prominent function for the commercial sector and, as the actual Bill made its way through parliament, the future position of commercial radio was becoming more and more attractive.

The Late 1990s: Threats and Opportunities

By January 1999 John Birt was approaching the end of his tenure as BBC Director General,[67] and in an attempt to bequeath a lasting legacy to the corporation, he initiated a 20-year strategy preparing the BBC for the digital future. Birt himself was becoming increasingly committed to the idea of internet radio and wanted the BBC to embrace this platform. If BBC Radio was to have any future at all, Birt argued it would have to develop a presence in that space (Birt 2002, 469). The upshot of Birt's analysis and insistence was a 'reining in of DAB ambitions within BBC Radio' coupled with a 'huge effort to develop a solid internet presence for the BBC's national stations' (Nelson 2003). For most of the established UK radio players, internet radio was seen as an important conduit in any bid to reach as many listeners as possible, but unlike Birt, the major commercial players remained focus on DAB as Capital Radio's Nathalie

---

[65] Virginia Bottomley, Heritage Secretary. *The Guardian*, 15 January 1996.
[66] Ibid.
[67] BBC Director General 1992–2000.

Schwarz[68] expounded: 'although the internet is important, in terms of scale of investment our prime emphasis is DAB.'[69]

Towards the end of the 1990s it is true to say the position of DAB was to fall in the ranking of the BBC's digital imperatives as the corporation adopted a policy of 'platform neutrality'; in other words, when it came to referencing digital platforms this meant availability on digital radio, digital television and the internet. Highlighting one particular platform went against BBC policy of the time. One can argue that following the departure of Liz Forgan as a champion of DAB it lost some of its currency under her replacement, Matthew Bannister,[70] who shared much of Birt's philosophy[71] and was able sometime after his tenure to reflect on digital radio as a 'costly dud.'[72]

Bannister's successor in 1999, Jenny Abramsky, was able to take DAB off the back burner and put it once again at the 'forefront of BBC Radio policy and indeed BBC policy in general' (Nelson 2003). When Greg Dyke became Director General in January 2000,[73] he freely admitted to possessing a poor understanding of radio and thus became supportive of Abramsky's desires to continue to pursue the DAB route (Dyke 2005, 176), trusting perhaps 'her superior knowledge and experience of the field of radio' (Nelson 2003). It is reasonable to conclude that Abramsky, with her passion for radio and long experience working within the BBC Radio division, was instrumental in ensuring the BBC would once again place DAB as its primary digital radio objective, thus reverting to what was its core policy only a short time previously. One can speculate how DAB might have prospered at the BBC had Birt's policy of internet continued for any longer than it did as surely the BBC would have lost ground against commercial opposition. Alas, further steps by Abramsky to launch new digital services renders such speculation unnecessary.

One of the first missions of the Commercial Radio Companies Association (CRCA) after its formation was to hold a DAB seminar to discuss the perceived significance of the technology for the commercial sector over the next 10 years.[74] Mark Story (Programme Director, Virgin

[68] Capital Radio Director of Strategy & Development.
[69] The New Pioneers. *The Guardian*, 18 August 2003.
[70] Director, BBC Radio 1996–1999.
[71] Bannister had played a central role in the *Extending Choice* project.
[72] *The Times*, 7 September 2001.
[73] BBC Director General 2000–2004.
[74] CRCA Seminar on DAB, 15 May 1996.

Radio) shared Virgin's enthusiasm for DAB, while Mark Flanagan (Managing Director, Fox FM) expressed the reservations of the ILRs who felt unhappy with the fact that DAB multiplex broadcasting might allow the bigger radio companies to take control of local broadcasting. This sentiment represents a division between the INRs and ILRs over DAB, the former becoming increasingly keen while the latter becoming increasingly cautious, fearing increased competition and particularly from within the ILR sector itself. In addition to the commercial radio companies, also present at the seminar was the transmission company NTL which had run the Birmingham trial in November 1995 and was able to report it had begun a longer-term trial in London in March 1996 featuring a number of the London commercial stations. Speaking of the launch, NTL's DAB Product Manager, Jon Trowsdale, stated:

> DAB for commercial broadcasters has taken on new momentum due to the Broadcasting Bill and the incentive created for existing broadcasters to transmit digitally in the years ahead. There are opportunities for new and innovative programme formats and flexibility for digital only services. We want to help commercial radio explore the great opportunities presented by DAB while saving them the worry of technicalities.[75]

The trial London multiplex was of course simply a trial as few people possessed receivers during this period and the fact that stations were rotated on the multiplex on a monthly basis[76] meant it was not treasured as anything more by those stations, but it does illustrate that the commercial sector was actively exploring how the technology would operate, how it would sound and what benefits it might bring.

The Broadcasting Act eventually received Royal Assent on 24 July 1996 and this represented a significant milestone in the history of the commercial radio sector's relationship with DAB. It set out the framework for the permanent licencing and regulation of DAB in the UK, allowing for one national multiplex to be operated by the BBC and one operated by the national commercial companies, as well as provision for at least 26 local and regional multiplexes which would serve both BBC and commercial services at that level. Since the existing INRs would have guaranteed places on the national commercial multiplex, that meant there would be room

[75] Ibid.
[76] Ibid.

for a further three additional services as well. It could be argued that what the Act offered in terms of the distribution of multiplexes was a parity of distribution for both the BBC and the commercial companies, so that in this post-FM era, the commercial sector would no longer be struggling against BBC dominance of frequencies but instead find itself finally 'on an equal platform' (Howard 2004) and be in a position to threaten the BBC's audience base.

The Act may be seen as a good thing for commercial radio as the INRs were guaranteed a slot on their own national multiplex and ILRs were to be offered their own highly attractive incentive. Those which decided to take their place on a local DAB multiplex would get an automatic eight-year extension to their existing analogue licence. This was a key induce-ment for commercial companies, so much so that while many remained sceptical towards DAB, the idea of incentivising commercial stations to join DAB was according to Stoller (2010, 280) a 'deal breaker,' but Stoller (ibid.) admits this effectively *forced* the entire sector to become supporters of DAB despite any misgivings. Regional multiplexes involved alliances between the big groups, for example; Talk Radio, Ginger Media Group and Clear Channel. Thus, we see the first steps of a commercial sector takeover by the big radio groups and the creation of a new com-mercial elite which would continue to dominate the sector for many years to come. It did however create a certain degree of tension as Rudin (2006) notes:

> The contrast in enthusiasm between small and large commercial companies for DAB was a feature from the time that it was being fully tested and promoted as a serious new technology which the industry had to come to terms with.

Another important dimension of the Act which facilitated the emer-gence of the big radio companies was the relaxation of media ownership limits which the government hoped would create a more competitive British media industry. This intention applied as much to the BBC as well, as it was now encouraged to boost its commercial activities. This new trend towards 'marketisation' (Levy 1999, 32) would completely alter the shape and strength of commercial radio as big groups would come to dominate the sector in an overt battle against a more commercial BBC. This substantial deregulation of media ownership meant that within ten years two companies, Emap and GCap, would come to own 60% of all commercial radio listening, but did it have any influence on the commercial sector's decision to embrace DAB?

Certainly the government perceived that a deregulated market would provide security for the commercial sector to take the step, and at the end of 1995, Virginia Bottomley was able to inform the Radio Authority of how she understood the strength of feeling that 'before investing in DAB, radio companies would need a greater security of tenure than was afforded by current licences'[77] and pointed to the forthcoming Broadcasting Bill[78] which would rectify that. This had been a particular bugbear for the most successful INR, Classic FM, whose licence was due to expire in five years and who actively sought some form of long-term stability if it is was to invest in digital technology. Quentin Howard, who was head of engineering at GWR at the time, and thus responsible for that company's decisions on technology, believes DAB could only have become a possibility in the commercial sector as a result of the emergence of big players in the market (including his own) who had the 'capacity to take the risk' and who ultimately believed DAB provided a key 'attraction in building further market positioning against the BBC' (Howard 2004). Stoller (2010, 161) implies the commercial sector played a clever game here and embraced DAB simply as a means to further its own objectives to secure its place in the radio market:

> As was to be the case throughout the creation of digital radio policy and legislation, the commercial radio industry was chiefly concerned with how far appearing willing to adopt the new transmission technology would be a useful bargaining chip as they pressed for de-regulation for their analogue services.

This suggests that the commercial sector was not wholly committed to DAB but instead saw that signing up for the new technology, by agreeing to government incentives, was in reality merely a means to attempt to assert further dominance in the analogue arena. If this was indeed smoke and mirrors, then Stoller (ibid., 156) goes on to lament the strategy as he feels it 'forced the entire industry to become apparent supporters of DAB when some public commercial scepticism would have been useful.'

---

[77] *The Guardian*, 8 December 1995.
[78] Broadcasting Bill 1995, House of Lords, Bill 19, London HMSO.

It is fair to say the 1996 Broadcasting Act made it clear commercial radio was to have an important role to play in the consolidation of DAB in the UK. As Lax et al. (2008) point out, there were a number of key areas which would give the commercial sector greater prominence:

> The awarding of the multiplex operating licences to the existing big commercial radio groups represented a major shift from the BBC in favour of the commercial sector, particularly at national level where there would now be as many commercial stations as public service stations. The control of the multiplexes by the large radio groups, coupled with a relaxed regulatory regime would help give these groups competitive advantage. The inclusion of an automatic renewal of existing analogue licences helped persuade commercial stations to risk investment in DAB. Also, the regulator, the Radio Authority, was to take a lighter touch … decisions on which stations should be carried on a particular multiplex were left to the multiplex operators and these decisions were made on a commercial basis.

The national commercial digital multiplex was advertised on 21 March 1998. There was only one application and it came from Digital One, a company perhaps unsurprisingly heavily dominated by GWR and which reflected what had become a passion for digital radio from GWR stalwarts Ralph Bernard[79] and Quentin Howard[80] who had built a reputation among those within the commercial sector as passionate advocates of DAB.[81] Howard in particular became recognised within the industry as a 'digital evangelist'[82] who was convinced of the future of digital radio:

> All media is going digital … to think that analogue will still be able to hold its own in ten years time is unrealistic, it won't be able to compete.[83]

Bernard and Howard became figureheads for DAB and were driven as much by the base economic needs of the companies as by the sound quality and extended services arguments which they often presented to commercial partners. Howard identifies 1997 as the year when his own

---

[79] Chairman, GWR.

[80] Chief Engineer, GWR.

[81] RadioCentre Roll of Honour revealed. *Radio Today*, 3 July 2013.

[82] *Commonwealth Broadcaster's Handbook and Directory* 2006, 31. London, Commonwealth Broadcasting Association.

[83] Quentin Howard: The Changing Face of Radio. *Director Magazine*, March 1999.

company GWR finally bought into DAB and believes it was his company which spearheaded the commercial sector's assault around this time. As he points out, other big companies, such as Capital in London, had told him they were content with their FM signal and the existing BBC London trial had not appeared to offer any viable threat (Howard 2004), but according to Howard:

> They were missing the point, it was about more choice, it was about more channels for your portfolio but what they really misunderstood was that it was about territory, it was about achieving more ground and that was important, particularly gaining ground on the BBC. (Howard ibid.)

It was not surprising therefore that GWR would be interested in bidding for the first national commercial multiplex once it was advertised by the Radio Authority on 24 March 1998. In the event, there was only one bid from Digital One, a consortium consisting of GWR and NTL. Digital One launched on 15 November 1999. Howard at this point became Chief Executive of Digital One and it is useful to consider his and the company's role in driving digital at this point. Howard (ibid.) admits as soon as Digital One was awarded the licence in October 1998, he set out on a mission to:

> Build a business model for commercial radio, sell capacity on the multiplex, create a raft of stations that would appeal to consumers and fundamentally create a marketing proposition.

It would appear Digital One wished to employ commercial principles in constructing its own model for DAB which would exploit DAB's technical advantages in order to take advantage of the market opportunities it presented. Howard (ibid.) is adamant that securing a primary market position for the commercial sector was of great importance but admits that in attempting to achieve this he decided to approach the BBC:

> One of the first things I did was to go and knock on the BBC's door and say, look, we should do this together … as the commercial company we have got the freedom do try things that you can't and if we pool together maybe we can both help make DAB work.

Although a relationship of sorts had existed between the BBC and the commercial sector in regard to DAB, Howard's statement suggests that it was Digital One who took the first step in promoting that relationship to a different level and most importantly narrowing the number of participants from the BBC and disparate entities such as the Radio Authority and a myriad of companies, to solely the BBC and Digital One, two entities who shared similar positions in the digital domain. Howard (ibid.) claims this worked because as he told the BBC:

> You have got the muscle, you have got the name, we are the upstarts but that means we can take the risks and see what can work and therefore together we should be able to build DAB successfully for the UK.

Digital One's approach to the BBC culminated in the first manifestation of formalised collaboration in the form of a cooperation agreement which was drawn up between both parties and which laid down parameters for the pooling of resources and research.[84] Howard claims it was this which would lead to the ultimate formation of the Digital Radio Development Bureau (DRDB)[85] in 2000, which became an incarnation of how the relationship would operate and one which Starkey (2008) believes:

> Provided the clear focus and coordinated strategy necessary for DAB to gain such a strong foothold in the UK market.

The timing of Howard's approach was fortuitous as it coincided with the appointment of Jenny Abramsky as the new BBC Director of Radio in January 1999 who was about to embark on a process of proselytism of DAB within the BBC itself and therefore for the BBC, a degree of support from the commercial side may have been deemed providential.

THE EARLY 2000S: SETS AND SERVICES

After the BBC and Digital one had joined forces, Howard (2004) admits there was still a concern within the commercial sector that there was a real danger of DAB not happening and that someone needed to intervene in

---

[84] Also cited in the BBC's submission to the DCMS Review of DAB, October 2004.

[85] DRDB was the radio industry trade body for DAB digital radio. Established in April 2000, it was funded and supported by the BBC, Digital One and other multiplex operators and the CRCA. Its task was to ensure DAB digital radio's wide accessibility and swift adoption in the UK alongside consistent and effective marketing. It was incorporated into the newly formed Digital Radio UK in 2009.

the market for receiving sets as this was where the ultimate solution could be found. Existing sets were cumbersome and expensive, and manufacturers maintained their reluctance to produce them at great volume as they were deemed to be a product with little profit potential. As Director of Engineering at GWR, Howard says he and his team identified that the main technical problem with the first generation of DAB sets was the silicon chip required for the sets, which was extremely expensive, thus forcing up the final retail price. GWR set about trying to develop a second generation chip which would be cheap enough to bring down the price substantially and also aid in altering the size of sets. The first portable DAB radio set for under £100 eventually became available in the UK in 2002 and was 'a direct result of GWR's financial input' (Rudin 2006). The Pure Evoke 1 was introduced by Videologic in July,[86] retailing at £99, thus breaking the £100 barrier. Manufacturers also felt they had crossed the DAB Rubicon with Pure confident that:

> At this price point the superb sound quality and content of DAB digital radio is finally accessible to the mainstream of UK consumers … the pent up demand for low cost, high performance DAB products can now be satisfied with a sustainable sub-£100 product.[87]

The commercial radio sector expressed its own joy at the arrival of the new £99 DAB receiver, declaring that it marked the beginning of:

> … exciting times for DAB digital radio. The Pure Evoke 1 is the first in a series of new, affordable DAB products that we expect to significantly push forward the DAB market during the second half of 2002.[88]

With the Pure Evoke 1 managing to stimulate the market and other manufacturers now joining the scramble to provide sets, one would guess that sales might begin to reflect this new-found market confidence. Alas, the most lucrative period in the following months came over Christmas

[86] Videologic is a British semi-conductor R&D and licencing company which became a division of Imagination Technologies in 1999 http://www.imgtec.com/. Pure is a British consumer electronics manufacturer http://www.pure.com/about/ and is a division of Imagination Technologies.

[87] Pure Press Release, 1 July 2002. http://www.pure.com/press-release/videologic-announces-first-pound99-dab-digital-radio-292/.

[88] Ian Dickens, Chief Executive, DRDB. DRDB Press Release, 1 July 2002.

2003 when 176,000 sets were sold,[89] a disappointing figure when compared to the initial investment costs. By July 2004, almost two years after the launch of the Pure Evoke 1, almost two years after the launch of the new BBC national services and five years after the launch of the national commercial multiplex, 682,000 sets had been sold and this was expected to reach the one million mark by the end of the year[90] which was significant for the radio companies but not so much for the manufacturers who, in an attempt to boost sales, began to divert production from single platform sets (offering DAB only) to dual platform sets (offering both DAB and FM)—this would provide a safety net should DAB fail to develop as a platform on its own.

GWR's decision to sign a deal with Imagination Technologies in March 2001 to invest an estimated £3 million towards making the chip was risk laden but also proved propitious, and Howard (2004) believes if they had not done so at that particular time then DAB may never have taken off. Stoller (2010, 288) in his work on DAB and the commercial sector overlooks GWR's specific investment in the manufacturing side but instead highlights how the DRDB also provided seed-corn funding towards the development of a third generation chip which he says eventually 'lifted the market from virtual stagnation to modest niche availability.' In terms of DAB landmarks, the creation of the new BBC services in 2002 was certainly one but the development of the chip that launched the £99 set in the same year was certainly another as Howard (2004) states:

> I am absolutely convinced, if we had not got the chip we would be looking at a market which would never have exceeded a maximum of 200,000 DAB sets and a price that could never have fallen any lower than £199 per set.

On the whole, 2002 was in general terms a good year for DAB; it saw the launch of the first sub-£100 set and also saw the introduction of new BBC services—thus we see content perfectly complementing technology and vice versa. Although sales of sets may not have been as high as expected, it still represented an increasing take-up of the technology among consumers. The potential of new services to entice both consumers and manufacturers had been considered by the BBC for some time and the

[89] Gfk research on behalf of DRDB, December 2003.
[90] Digital Radio Take-up to Break 1m Mark by End of 2004. *Campaign*, 29 September 2004

idea began to appear even more attractive if it might give some headway against the commercial sector. As the BBC surmised, since there was no mention of new services at all at this point in discussions with commercial representatives under the auspices of the UK DAB Forum, this offered a 'window of opportunity to announce new services with a lower than usual risk of political flack from commercial radio.'[91] Simple simulcasting of existing analogue services would mean that a BBC digital multiplex would not necessarily offer any attractiveness for listeners, so the BBC announced its new digital radio service plans in October 2000 and submitted them to the Department for Culture, Media and Sport (DCMS) in January 2001.

In a speech to the Radio Festival in July 2001, Greg Dyke outlined his position on radio, describing how he envisaged the future of a medium he had now become familiar with:

> Analogue radio is no longer able to provide the industry with a way of growing their business or provide listeners with more choice … outside of the media I have yet to meet a single listener with a digital radio … we've invested over £30 million of licence payers' money in digital radio and to be frank we've seen very little in return. So why are we so keen on digital radio? … it's clear the internet can't replace broadcast radio for our core UK networks … digital radio is the only technology on offer that allows radio to grow. But there's still a long way to go before digital radio ceases to be just a technology and becomes a consumer product. Just like the commercial sector we're taking quite an expensive punt that digital radio will be central to the way radio is consumed in the future. But for it to succeed we have to find ways to get the price of receivers down … We are prepared to invest more. We can't use public money to directly subsidise receivers but we can invest in new, high quality services.[92]

Dyke's speech is interesting as it addresses a number of issues concerning the BBC's position with regard to DAB. It consigns analogue radio to the dustbin of media history, thus finally putting pay to any assumptions that FM may persist as a preferred model or that John Birt's idealised model of internet radio could represent the future of radio broadcasting. Dyke's statement unequivocally reinforces the BBC's commitment to

---

[91] DAB Update & Issues for Consideration. Paper for Radio Directorate meeting, 7 February 1996. BBC WAC: E120-014 Part 2 Digital Audio Broadcasting—Project Group.

[92] Public Service in a Digital World. Greg Dyke speech to the Radio Festival, 10 July 2001. BBC Press Release.

DAB. At the same time however it highlights the fact that DAB penetration has not been of sufficient viability to meet the BBC's significant investment thus far. It would seem that the BBC may not wish to devote more public funding leverage towards DAB, but Dyke believed that the prohibitive price of receivers remained the last obstacle to DAB and therefore an investment in new services might help overcome this obstruction. As an introduction to his speech, Dyke was not embarrassed to admit that as a 'television man' his understanding of radio was pretty limited when he took up the job of Director General almost two years previously, yet his decision to continue with DAB and invest even more licence fee money towards it seems bold and categorical.

It can be argued that's Dyke's mission was in fact driven from within the BBC's own Radio division. Nelson (2003) recalls that Dyke indeed did not possess a great knowledge of radio and 'took his lead on the subject from the existing BBC Director of Radio, Jenny Abramsky.' Dyke himself admits Abramsky 'should get most of the credit for turning British radio digital' (Dyke 2005, 176) and he believes she was correct in her persistence:

> She persuaded, cajoled and threatened everyone inside the BBC to support the plan to develop a series of new BBC digital radio services. And it was the arrival of those services that finally persuaded the radio manufacturers to start producing digital radios. (Dyke ibid.)

While it may be logical to conclude that Abramsky was the driving force within the BBC to provide the new services which would then aid in the uptake of DAB, it could also be argued that in effect it was both Abramsky and Dyke who brought about the new crucial dimension to digital expansion, namely new digital specific channels. While Abramsky's motivations were centred on her desire to promote the medium of radio, Dyke's digital strategies were, as Born (2004, 486) notes, based on completely different assumptions:

> Birt's instinct had been to deploy digital for polar ends: on the one hand commercial expansion, and on the other pure public service purposes by attempting to reach underserved communities. Dyke adopted a similar philosophy but took a more mainstream if resolutely public service orientation, one peppered with pragmatism. To the Reithian mission to inform, educate and entertain, Dyke added 'connect'.

Dyke's digital proposals ended up including a portfolio of five new digital radio networks and four complementary free-to-air digital television (DTV) channels, so this was a digital package which included radio as much as television. The five national radio networks submitted included Radio 5 Live Sports Extra (a continuation of 5 Live Sports Plus, a part-time network exploiting the BBC's sports rights to the full), as well as the Asian Network (an existing regional service which had been providing dedicated Asian programming across the Midlands). Three brand new services were also planned and at this stage went by the simple nomenclature of Networks X, Y and Z. Network X would be an off-shoot of Radio 1 and provide 'cutting edge black music.'[93] Network Y would provide music for an audience caught between Radio 1 and Radio 2,[94] while Network Z would provide a mix of 'intelligent, fun, vintage and new shows for adults and children.'[95]

The BBC received government approval to launch all its digital services on 13 September 2001, which covered its range of five digital radio channels and three[96] of its four proposed television services.[97] The BBC described this point as 'putting an end to 8 months of uncertainty and opening up a new era of digital development.'[98] thus considering it a turning point in its digital history and vindication of its confidence in DAB. Dyke saw this as an enormously exciting time for the BBC as a whole, both television[99] and radio:

> There have been times in the BBC when there has been expansion, this is one of them … A lot of money is going into production to make a lot of new services … We are seeing the biggest increase in programming expenditure in the history of the BBC. Between 2000 and 2002 we will will have increased spending by £450 million.[100]

[93] New Television and Radio Get Go-Ahead. *Ariel Digital Special*, 14 September 2001.
[94] Ibid.
[95] Ibid.
[96] BBC 3's approval by Culture Secretary Tessa Jowell would have to wait a little longer when its revised proposals were eventually accepted.
[97] BBC Children's 1 (later named CBeebies) for children under 6; BBC Children's 2 (later named CBBC) for children 6–13; BBC Choice (later named BBC 3) for a 16–34 audience and BBC 4 for cultural and factual programming.
[98] New Television and Radio Get Go-Ahead. *Ariel Digital Special*, 14 September 2001.
[99] For more analysis of digital television in Britain, see Starks (2007).
[100] New Television and Radio Get Go-Ahead. *Ariel Digital Special*, 14 September 2001.

Abramsky, perhaps unsurprisingly, focused on the importance of radio and the new radio services:

> I am really heartened that the secretary of state emphasised the important role that BBC radio has to play in driving digital take-up and that she saw our services as distinctive and adding to our public service remit.[101]

Culture Secretary Tessa Jowell shared Abramsky's confidence in the key role radio in particular would play in digital expansion in the UK. In a speech to the Royal Television Society on the day of the announcement of the new services,[102] she expressed her hope that the BBC could now play to its historic PSB strengths in promoting Britain's digital environment:

> More than any other institution the BBC has the capacity to build the the digital radio market. It is proposing to offer distinctive and attractive new services to audiences not currently well served across the UK. A strong BBC presence on digital radio should increase listeners and further encourage manufacturers to bring new products on the market at prices that people can afford.

Jowell recognised that digital take-up of radio had been painfully slow despite the fact the BBC had been offering a service for the past six years. Less than 40,000 sets had been sold during this time and the exorbitant price tag of £300 was certainly a major disincentive. Jowell's holy grail was a set retailing at £99 and she hoped such a set would be available for the Christmas market of 2001.[103] Commercial operators involved in set manufacture did not believe this was feasible and Jowell was probably being over-optimistic as although the new BBC services had been approved, they would not roll out until 2002. By this time Network X would become BBC 1Xtra,[104] Network Y would become BBC 6 Music[105] and Network Z becoming BBC 7.[106] A curious aspect of the new services was that the word 'radio' was missing from the station names, this representing a con-

---

[101] Ibid.

[102] Public Service Broadcasting in the Digital Age and the New BBC Services. Speech by Tessa Jowell to the Royal Television Society, Cambridge, 13 September 2001.

[103] Ibid.

[104] BBC 1Xtra launched 16 August 2002.

[105] BBC 6 Music launched 11 March 2002.

[106] BBC 7 launched 15 December 2002. Relaunched as BBC Radio 4 Extra on 2 April 2011.

tinuation of the BBC's emphasis on 'platform neutrality' or in other words, multi-platform presence.

Rudin (2006) argues that by 2004, two years after the launch of its new services, the BBC had accepted that improved sound quality had been far less important than the increase in programme choice for DAB to be a success. Rudin bases his claim on the BBC's submission to a DCMS review of DAB performance where the BBC was able to claim:

> The extension of listening choice was fundamental to driving take up of DAB, with wider choice consistently cited by digital radio owners as the main reason for buying sets.[107]

But surely the BBC had accepted the fact that new services were more important than sound quality quite a few years before this? It was after all the reasoning behind the desire to create new content in the first place, and Rudin overlooks the fact that Liz Forgan had first identified the importance of new content even as far back as 1996.

The BBC's decision to pursue DAB was proven to be correct and its decision to provide new services was deemed to be auspicious, as more new national BBC services had been launched over the course of 2002/2003 than at any point in the corporation's history. Chairman Gavyn Davies was able to highlight the success of the new services in the BBC Annual Report for 2003. BBC Radio's new digital stations had added 400,000 new listeners, reaching out to a record 53.5% of the total radio audience and earmarked BBC 7, which in its first week achieved 920,000 page impressions on its website, the highest launch figure for any of the new services.[108]

Sets and services were the two necessary elements required for DAB to be at least in a position to succeed. The steps taken by GWR (and then Digital One) and the BBC in helping to provide sets and services may be seen as singular actions, but once both were in place then it would seem inevitable that the two erstwhile enemies might seek to instigate a policy of cooperation in order to achieve the same goal. A policy of cooperation had of course already been implemented under the auspices of the DRDB where the BBC and the CRCA were sharing conference platforms and collaborating on information or going on fact-finding missions. The DRDB

---

[107] DCMS Review of DAB Digital Radio—The BBC Submission, London, 2004.
[108] BBC Annual Report 2002/2003. BBC Publications, July 2003.

also launched a series of marketing campaigns across national, regional and local analogue commercial radio stations through 2002 explaining DAB and using the BBC's John Peel[109] as a trusted voice who could reach across the generations (Chignell and Devlin 2006). The BBC meanwhile launched its own series of promotions for its new services as they launched over 2002 with trails appearing across its analogue radio stations and television channels, again attempting to educate audiences about what DAB could offer. A big step in this educative strategy by the BBC and the commercial sector was the unveiling of a logo and brand identity for DAB in April 2002 which was to be used on all literature and point-of-sale material for DAB digital radios, as well as advertising across all media.[110] While the DRDB was striving to promote DAB for both the BBC and all the commercial sector, there was a tangible strategy of cooperation continuing separately between the BBC and Digital One; in fact the relationship became very close with visits between Digital One and the BBC's Audio and Music Interactive department becoming commonplace (Nelson 2003; Howard 2004) and geared at promoting knowledge and acceptance of DAB among staff.[111] A shared strategy between the BBC and Digital One seems obvious since both had made such a huge commitment to DAB, but the BBC may have been somewhat more cautious of the CRCA as it had already lobbied the DCMS to restrict the BBC's new services, fearing the impact they would have on new commercial outlets and citing as an example how BBC 7 might impact on Oneword[112] which was launched by Unique in May 2000. It was evident that the BBC and Digital One had decided to pursue their own micro-strategy of DAB promotion, believing perhaps that they were the technology's true champions.

CONCLUSION

The BBC may be seen as the UK's digital radio champion for the most part of the 1990s, particularly in its involvement at the early stages of the technology's development and then in running the UK's first DAB trials

[109] John Peel: the New Voice of Commercial DAB Digital Radio. DRDB press release 18 January 2002 http://www.ukdigitalradio.com/news/display.asp?searchnews=&year=2002 &id=106.

[110] DRDB Reveals Identity for Digital Radio Technology. *Marketing magazine*, 11 April 2002.

[111] The author also recalls the numerous meetings extolling DAB and offers of discounted DAB sets available to BBC and commercial sector staff.

[112] A national digital speech network which remained on air until 2008.

and in its attempts to proselytise the manufacturing industry. The BBC's reticence towards DAB and focus on internet radio at the end of the 1990s, coupled with the subsequent availability of a national commercial multiplex, meant that the commercial sector then acted as digital champion for a critical period. An important role for the BBC then re-emerged in the early 2000s with the launch of its new services. Hendy (2000b) succinctly concludes that it would appear that large national public service broadcasters, with their strategic ability to invest in long-term research and without the need to deliver an immediate return of large audiences to advertisers, have been central to the early development of DAB, with commercial operators following in their wake. This is apparent in the UK model and Hendy's (ibid.) conclusion that 'digitalization is thus cast as the end-game of deregulation' allows to consider that the commercial sector too can take a lead when appropriate and indeed the role of Digital One in contributing towards the reduction in the price of sets is an example of this. However, the arrival of new services and cheaper sets did not necessarily mean DAB was now set to become the standard for radio listening.

At a meeting of the Radio Academy in 2003,[113] focusing on DAB, it was evident that even among the 'radio people' in attendance there was a feeling that despite the significant progress that had been made, DAB was still not achieving the results that had been hoped for. The Session Chairman, Roger Bolton, asked those assembled how many had a DAB radio, to which only about 60% had purchased one. If this was the ratio within the industry then it would appear that DAB was still a fledgling medium. Some contributors at the meeting were concerned about this. Ron Coles from Saga[114] was adamant more sets needed to be sold if companies like his were to survive, while Andrea Kilbourne from Emap[115] claimed that digital television platforms and particularly Freeview,[116] which launched in September 2002, had helped more in promoting their digital radio brands than DAB. There was a feeling among some commercial

[113] Radio Academy monthly meeting, May 2003. Off Air (Radio Academy newsletter, Autumn 2003).

[114] Saga operated Primetime Radio on the national multiplex as well as local Saga Radio digital services.

[115] EMAP operated 7 local DAB multiplexes and owned 40 UK and Ireland local commercial radio stations. Now part of Bauer Radio.

[116] Free-to-air digital terrestrial television service. It is a joint venture between the BBC, ITV, Channel 4, Sky and transmitter operator Arqiva (see Given and Norris 2010).

players that DAB was still not going to happen and Stoller (2010, 288) encapsulates that sentiment when he states: 'DAB was not working. What rescued digital radio was digital television.'

While the smaller commercial stations were becoming more hesitant on the grounds that few receivers had been sold (Lax et al. 2008), other elements were pointing the finger at the BBC for the initial slow performance of DAB. Some commercial operators took umbrage at the BBC's decision to launch its new services and blamed the corporation for introducing a wedge of unfair competition. Interviewed for a special issue of the BBC's *Ariel* in 2001,[117] Patrick Berry of Choice FM[118] objected to Network X which he feared would undercut the digital version of his own station. Meanwhile Shujat Ali of Manchester's Asian Sound had similar concerns regarding the BBC's intention of making its Asian Network available nationally on DAB, which he claimed would 'suffocate' his stations.[119] The BBC rejected these criticisms claiming it had evidence that the commercial industry as a whole was in favour of the BBC's plans for new services on the grounds that they would encourage the take-up of digital receivers.[120] This raises the interesting question of whether the BBC and Digital One were on a course of DAB promotion together which left the other commercial operators out on a limb. When asked about this, Howard (2004) replied that the compelling argument for Digital One was to match the BBC in terms of a national multiplex as this would create a level field upon which to compete with the BBC, a luxury the commercial sector never had before. It could not therefore very easily challenge the BBC's own multiplex or the genres of stations on it, all it could do was to compete and attempt to continue the commercial sector's recent period of supremacy against its arch enemy.

The commercial sector was still maintaining its healthy status as the new millennium began, so one can understand Howard's desire to capitalise on this position. By 1999 local commercial stations were capturing nearly 40% of all radio listening between them and British listeners were more likely to be tuned into a local commercial station than any other service.[121] In 2000, 66% of the UK adult population was listening regu-

---

[117] New Television and Radio Get Go-Ahead. *Ariel Digital Special*, 14 September 2001.
[118] A 24-hour black music station covering South London since March 1990.
[119] New Television and Radio Get Go-Ahead. *Ariel Digital Special*, 14 September 2001.
[120] Ibid.
[121] RAJAR, Q1 1999.

larly to a commercial radio station[122] and by 2001, 77% of all listening to
UK local radio was captured by the commercial sector[123] and the sector
was proving profitable ground for advertisers with commercial radio revenue from advertising reaching £577 million between June 2000 and June
2001.[124] But it was not without its detractors, with the greatest criticism
directed at the fact that most stations offered the same fayre of Top 40
hits. Irvine (2000) believes one of the sector's main critics in this regard
was actually its own regulator, the Radio Authority, which in a consultation on a new White Paper in 2000[125] advised the government to establish
a third sector of radio distinct from the BBC and the existing commercial
stations. It called this tier 'Access Radio' and promoted it as 'a means to
assist in education, social exclusion and experimentation and thus extend
the diversity of radio services.'[126] The conclusion from within the commercial sector was that the Radio Authority did not think it was fulfilling
its goals (Irvine ibid.).

Stoller calls Section IV of his book *Victory of the Commercial Model
1990–2003* and it is an apt title. It was the era when the commercial sector
came to the fore. In analogue radio terms it caught up with the BBC,
equalled it and then overtook it in 1995 and maintained its position of
strength. As far as digital radio is concerned, I suggest an amendment to
this title, instead *Victory of the Commercial Model 1997–2002*. Before 1997
one has to admit that the BBC was the DAB champion, it helped develop
the technology and implemented a sustained trial, indeed those in the
commercial sector concede this fact (Howard 2004). After 2002, when
the BBC launched its new digital only services, one could argue we had
moved back into equilibrium with the BBC gradually in the ascendancy.
However, in the intervening period there is no doubt the commercial sector was the prime driver of DAB, and a number of factors were key here.
Between 1998 and its disbandment in 2003, the Radio Authority had
awarded 45 local multiplexes offering more than 270 programme services
(Stoller 2010, 286). It also awarded one national multiplex and the advertising of that multiplex was crucial, although there was not a great deal of

[122] RAJAR, Q2 2000.
[123] RAJAR, Q2 2001.
[124] Commercial Radio Revenues. Radio Advertising Bureau, 2001.
[125] White Paper: A New Future for Communications. DTI/DCMS 2000. London, HMSO.
[126] Radio Academy response to White Paper: A New Future for Communications. Radio Academy, February 2001.

competition for it—in fact it only had one bidder, Digital One—but what is important is that it provided an opportunity for those within the commercial sector who believed in DAB to pursue it in their own manner. This step ultimately led to another factor, namely, that of the chip which Digital One had invested so heavily in and which brought down the price of receivers which had been a major obstacle. A final factor was the diminishing enthusiasm within the BBC which occurred around the same time. The BBC's commitment was foundering due to an insouciance at the highest echelons within the organisation and if the commercial sector had not bolstered a flailing DAB at this point, it is reasonable to question if it might have survived at all. Digital One, having secured the national multiplex in 1998, embarked upon a sustained programme of investment that ten years later had still failed to produce a profit while the manufacturers had already achieved the economies of scale required to produce DAB sets economically (Starkey 2008). But its role in sustaining DAB through the late 1990s should not be overlooked and allows us to conclude that during that crucial period, the commercial sector was the DAB champion in the UK before the BBC was in a position to launch its new services, thus providing the two-pronged assault that would be necessary to take DAB to the next level.

REFERENCES

Ala-Fossi, M., Lax, S., O'Neill, B., Jauert, P. and Shaw H. "The Future of Radio is Still Digital – But Which One?: Expert Perspectives and Future Scenarios for Radio Media in 2015." *Journal of Radio and Audio Media* 15, no. 1 (2008).

Berlemann, L. and Mangold, S. *Cognitive Radio and Dynamic Spectrum Access.* Chichester: John Wiley & Sons Ltd, 2009.

Birt, J. *The Harder Path.* London: Time Warner, 2002.

Born, G. *Uncertain Vision: Birt, Dyke and the Reinvention of the BBC.* London: Secker and Warburg, 2004.

Chignell, H. and Devlin, J. "John Peel's Home Truths." *The Radio Journal: International Studies in Broadcast and Audio Media* 4, no. 1, 2 and 3 (2006).

Cornell, L. (BBC DAB R&D, 1995–2000; Head of BBC Digital Platforms, 2001–2004; Chairman of World DMB Technical Committee, 2001). Interview with author, 7 August 2003.

Crisell, A. *Understanding Radio.* London: Routledge, 1994.

Dyke, G. *Inside Story.* London: Harper Perennial, 2005.

Given, J. and Norris, P. "Would the Real Freeview Please Stand Up?" *International Journal of Digital Television*, 1, no.1 (2010).

Hendy, D. *Radio in the Global Age*. London: Polity Press, 2000a.

Hendy, D. "A Political Economy of Radio in the Digital Age." *Journal of Radio and Audio Media* 7, no. 1 (2000b).

Howard, Q. (Director of Engineering, GWR, 1982–1987; Chief Executive, Digital One, 1998–2008). Interview with author, 1 May 2004.

Irvine, N. "Commercial Radio: Serving UK Communities." *Cultural Trends* 40, no.1 (2000).

Lax, S. The Prospects for Digital Radio: Policy and Technology for a New Broadcasting System. *Information, Communication and Society* 6, no. 3 (2003).

Lax, S. "Digital Radio and the Diminution of the Public Sphere." In Butsch, R. (ed) *Media and Public Spheres*. Basingstoke: Palgrave Macmillan, 2007.

Lax, S., Ala-Fossi, M., Jauert, P. and Shaw, H. "DAB: the Future of Radio? The Development of Digital Radio in Four European Countries." *Media Culture and Society* 30, no. 2. (2008).

Lax, S. *Media and Communication Technologies: A Critical Introduction*. Basingstoke: Palgrave Macmillan, 2009.

Lax, S. "The Failure of a 'Success Story': Digital Radio Policy in the UK." *Australian Journalism Review* 36, no. 2 (2014).

Levy, D. *Europe's Digital Revolution: Broadcasting Regulation, The EU and the Nation State*. London: Routledge, 1999.

Nelson, S. (Controller BBC Radio and Music Interactive, 2000–2006). Interview with author, 27 April 2003.

O'Neill, B. and Shaw, H. "Radio Broadcasting in Europe: The Search for a Common Digital Future." In O'Neill, B., Ala-Fossi, M., Jauert, P., Lax, S., Nyre, L. and Shaw, H. (eds) *Digital Radio in Europe: Technologies, Industries and Cultures*. Bristol: Intellect, 2010.

Poole, I. *Basic Radio Principles and Technology*. Oxford: Newnes, 1998.

Rudin, R. "The Development of DAB Digital Radio in the UK: The Battle for Control of a New Technology in an Old Medium." *Convergence: The International Journal of Research into New Media Technologies* 12, no. 2 (2006).

Scannell, P. "The Ontology of Radio." In O'Neill, B., Ala-Fossi, M., Jauert, P., Lax, S., Nyre, L. and Shaw, H. (eds) *Digital Radio in Europe: Technologies, Industries and Cultures*. Bristol: Intellect, 2010.

Starkey, G., 2008. "The Quiet Revolution: DAB and the Switchover to Digital Radio in the United Kingdom." *Zer* 13, no. 25 (2008).

Starks, M. *Switching to Digital Television: UK Public Policy and the Market*. Bristol: Intellect, 2007.

Stoller, T. *Sounds of Your Life: The History of Independent Radio in the UK.* New Barnet: John Libbey. 2010.

Street, S. *A Concise History of British Radio 1922–2002.* Tiverton: Kelly Publications, 2002.

Tacchi, J. "The Need for Radio Theory in the Digital Age." *International Journal of Cultural Studies* 3, no. 2 (2000).

Thomas, M. "Commercial and Local Radio DAB." *Audio Engineering Society, Conference Paper. DAB-14, The Future of Radio, 1 May 1995.* London: Audio Engineering Society, 1995.

CHAPTER 8

Conclusion

Any history of radio in the UK must surely celebrate the medium's success, particularly as we approach the 100th anniversary of the first 2LO broadcasts in only a few years' time. That history must also surely appreciate the roles played by the actors within the industry over this period and account for their roles in helping sustain the medium into the long term. In doing so, it would also be appropriate to question the nature of the relationship between these actors in order to ascertain whether radio's success has been the result of joint enterprise or of singular actions based purely on self-interest. The aim of this book has been to allow us to appreciate the actions and consequences of the BBC and the commercial radio sector in creating the necessary environment for radio to endure nay flourish.

THE ROLE OF COMPETITION

For the BBC, it emerged from an initial cooperation among radio manufacturers before becoming a national entity enshrined with a PSB remit as the very nature of broadcasting took shape. Cooperation among manufacturers was a necessary pre-requisite to the evolution of an entirely new industry in order that it may establish and grow and to avoid a chaos of the airwaves. The move from company to corporation with a defined public persona was deemed necessary in order for a template of broadcasting to become enshrined in law, one that would uphold standards of taste in a

© The Author(s) 2018
JP Devlin, *From Analogue to Digital Radio*,
https://doi.org/10.1007/978-3-319-93070-1_8

new medium whose boundaries had not yet been marked. The BBC model proved successful in endearing the public to the medium of radio through programming and helped ensure the matching of content and technology to produce a service of public value. The emergence of competition in the 1930s was inevitable. Budding entrepreneurs were able to identify how other models of broadcasting had the potential to provide lucrative financial returns and so we witness the arrival of commercial radio, typified by the use of advertising to secure revenue and the production of populist programming to maximise that revenue—it was also typified by its illegality. Following the arrival of competitors in the 1930s, the BBC initially adopted a phlegmatic approach, secure in its own position as the country's sole legitimate broadcaster and only latterly embarking on a journey of reconciling its role as PSB guardian with the need to adapt to changes in audience demands if it was not to suffer from the effects of commercial opposition. Of this early period we can say that the formation of the BBC had the effect of creating stability in a fledgling industry and that it created a template of programming that testified to its PSB credentials. Indeed, this template would continue to form the basis of BBC output over the following decades and help bestow upon the corporation a sense of trust from UK audiences and assure its place as the national broadcaster. The commercial broadcasters of the time set in place an entirely different model of radio broadcasting, one that could engender wide appeal as a result of a lighter form of programming, unconstrained by regulation and characterised by a bold and brash persona, not afraid to take risks and ready to initiate or adopt changes in broadcasting style. This was a basic model which would pervade although in many different guises, but one which would open up the radio market and ensure for broadcasters it would remain competitive, and for audiences ensure it would remain healthy.

The Second World War may have saved the BBC from any further erosion of its position on the British radio landscape as the commercial stations were forced off the air. During the war years it was able to re-establish its place as the primary broadcaster and the very nature of the war ensured this role was both necessary and appreciated. The effect of the war however was to create a simultaneous desire among the public for both information and distraction in equal measure, so at this time we see the BBC attempting to widen its PSB remit to include a more populist form of programming. This would take on board concepts of broadcasting which had been the central features of pre-war commercial radio although without expanding fully into this domain. In preparing for the post-war environment, the BBC demonstrated that it accepted the strategies of the commercial broadcasters to a certain extent but was careful to insist on its PSB role as its main raison d'etre. The war years and early post-war era permitted the BBC to regain its monopoly position but it also allowed it to adopt some commercial characteristics in terms of programme output (as opposed to economics). Before the commercial sector would reappear, it had already left its legacy on British broadcasting into the long term and on its return it would drive even further change.

Pirate radio of the 1960s tapped into societal changes and changes in audience dynamics. It also came about at a time when certain voices were clamouring for separate radio services distinct from the BBC. The pirate stations certainly brought even greater pressure to bear on the BBC to engage with a certain tranche of listeners which it had largely failed to accommodate. Relying too much on its PSB persona, and an obsession with standards, had always been a failing of the BBC to some extent as this left it open to the vagaries of the broadcasting market which instead had an obsession with audience figures. Chasing audiences was something commercial radio was rather good at and something which the BBC generally played as catch-up when it became exposed to the risks a passivity in this area exposed. Such was the case with pirate radio which must be celebrated for initiating a model of music broadcasting that has sustained to the present, not only by its commercial successors but also by the BBC itself. Once again, in the 1960s the BBC would be saved by legislation which would enable it to re-establish as the pirates were shut down.

Legislation also had the effect of creating competition under the Sound Broadcasting Act 1972 and the Broadcasting Act 1990 with competition now standard on both a local and national level. Since 1972 the BBC has had to face a commercial sector whose position has been strengthened by successive legislation. It is reasonable to conclude that on many occasions, the BBC has had to react to competitors' actions in order to maintain a position within the radio market, and one could argue that an ambivalence towards audience focus was not completely overcome until the threat of legalised competition emerged in the 1960s. With a strong PSB remit and a guaranteed funding model, it is perhaps unsurprising that the BBC would chase standards rather than audiences and its function as an upholder of PSB deserves credit, particularly with regard to news broadcasting. Likewise, its support of the arts and endeavours to uphold quality is reflected in the successful, and in some cases groundbreaking, programming it has produced over the decades. One can argue that the BBC has at times been the pioneer of certain forms of radio such as local radio or national light entertainment or pop music services. On the level of legalised broadcasting this is indeed the case, but these were usually reactionary steps following on from steps already taken by its competitors who have often been responsible for initiating change to the very shape and style of radio broadcasting itself, often spearheading format and presentation changes which were then subsequently pursued by the BBC. One can argue that in the evolution of radio broadcasting, the BBC has played an important role as public service broadcaster and defender of standards of broadcasting. The commercial sector has played an equally important role in acting as an alternative to the BBC, particularly in serving populist demands, and that today's radio industry is heavily reliant on those steps made by commercial companies, legitimate or otherwise.

THE ROLE OF COOPERATION

The arrival of DAB represents a useful demarcation point in that it heralds the fostering of a spirit of cooperation which had little historical precedent. Before this, the BBC had been the dominant radio player for some time despite significant démarches from the commercial sector, although this began to falter after the introduction of legalised commercial broadcasting. The balance of power shifted around the same time as DAB emerged as a new technology, with a greater degree of equality typifying the structure of the industry, and it was this very similitude which helped provide the perfect launch pad for DAB which in turn instigated a novel spirit of cooperation.

The radio industry has largely been characterised by traditional notions of competition based on classical, Marxist descriptions of competition between firms which are deemed 'a warlike process' (Moudud 2013, 30) and this is evident throughout the history of radio in the UK, but it changed for a period due to the arrival of digital technology. With both the BBC and the commercial sector now on a more level playing field as a result of legislative change, and as digitalisation came to pervade the entire media industry, both parties accepted a degree of symbiotic cooperation as an essential short-term pre-requisite in order to ensure the successful penetration of a technology which had perceived long-term benefits for both parties. Hence, for a duration, the classic Marxist interpretation of competition is replaced by what Dimmick (2003) describes as a 'niche theory of competition' whereby the rise of a new medium (or in this case a new technology) competes with established media (or in this case an established technology) for consumer interaction, satisfaction and revenue. The consequences of competition in such an environment include possible displacement or exclusion, as a result, competition reverts to simple co-existence in order to facilitate a successful transition towards the new technology.

Although Dimmick (ibid.) employs the term 'co-existence' to describe collaboration in a competitive arena, I use the term 'cooperation' in this study as it implies a more active form of mutual undertaking and one which requires a proactive participative role by all parties. A number of important factors characterise the DAB era and inspired its cooperative dimension. Firstly, the BBC and the commercial sector became equal players within the radio industry after many years of BBC domination and competitors' attempts to redress this imbalance, and secondly, technological change meant that a new platform for radio was deemed essential by both the BBC and the commercial sector due largely to the fact that spectrum shortage on the traditional analogue platform[1] meant growth for the industry would be severely constrained. These factors of equality and necessity provided the foundations for cooperation as a means of ensuring the survival of radio as an industry at a time of momentous change.

For the BBC, its decision to pursue DAB was also based on its strong PSB remit. As Hendy (2000) points out, it has been large public service broadcasters with their 'strategic ability to invest in long-term research and without the need to deliver an immediate return of large audiences to advertisers' that have been central to the development of DAB. Indeed, the BBC has been able to invest in digital (radio, television and online) through the licence fee[2] and has played a prominent role in building digital Britain,[3] which in turn has led to changes to the BBC model through encroaching notions of alternative funding in order to aid its transformation into an 'international multimedia enterprise'[4] and one which was no

---

[1] The Future Management of the Radio Spectrum: A Consultative Document. Radio Communications Agency (an executive agency of the Department of Trade & Industry), March 1994. London, HMSO.

[2] The BBC spent £154 million (7% of licence fee income) on digital services in 1998/1999. BBC Annual Report & Accounts 1998/1999.

[3] Building Public Value: Renewing the BBC for a Digital World. June 2004. BBC Publications.

[4] White Paper. The Future of the BBC: Serving the Nation, Competing Worldwide, July 1994. (DNH 1994).

longer 'commerce averse'—particularly under Greg Dyke (Born 2004, 475). As Born notes, Dyke's approach was one of public service orientation 'peppered with pragmatism' (Born ibid., 486).

The commercial sector's 'increased prominence and market presence and desire to build on its success by creating even more stations' (Howard 2004) meant it was ready to embrace a platform that would allow this to happen. The sector knew it was making significant steps in finding appreciation among the radio audience and was, according to some reports, 'beating the BBC on both a national and local level in reflecting listener concerns and lifestyles.'[5] The commercial sector was however hesitant initially due to the weakness of DAB receiving set sales, which is why it invested so much in developing the chip technology that would drive down the price and after that the sector played a significant role in DAB investment, particularly in programming.[6]

Any degree of active cooperation between the BBC and commercial radio is not apparent in the years preceding DAB, in fact one can argue that relations between both parties were virtually non-existent and the relationship could be described as distant. As the radio market began to change and become in effect a duopoly, then the first manifestation of a coordinated approach came with the formation of RAJAR in 1992. As the changing face of the industry necessitated cooperation regarding audience measurement, then so did it provide the impetus for cooperation in order to promote digital technology. The creation of the DRDB in 2000 was the joint response by the BBC and the commercial radio companies to promote DAB digital radio's adoption in the UK and was a genuine effort by both parties to ensure a successful future for both. Under the auspices of the DRDB there was much toing and froing between Broadcasting House and the offices of the big radio companies with the singular aim of publicising DAB.[7] It also included joint missions to the USA[8] to examine the digital market there and to Japan[9] to attempt to persuade manufacturers of the potential of the UK DAB market, although these are examples of

[5] Media Futures. 1993. The Henley Centre, London.
[6] By 2008 the sector had injected an estimated €200 million into the development of DAB (*Radio Magazine*, Issue 825, January 2008).
[7] The author attended a number of these sessions.
[8] The Future of Radio: A Mission to the USA. DTI Global Watch Mission. September 2004.
[9] Mission to Japanese Radio Manufacturers. Simon Nelson (BBC Audio & Music Interactive) report to staff. March 2004.

actions which also included input from the UK government. The DRDB continues to exist in the form of Digital Radio UK although the days of close contact have long passed and the relationship is now much more formalised and centres around broader issues such as drawing up digital radio action plans for technical specifications of receiver sets and promoting further market penetration. But as Starkey (2008) notes, the DRDB provided 'clear focus and coordinated strategy which was necessary for DAB to gain a foothold in the UK.' Whether it has been successful or not is another matter.

A SUCCESSFUL STRATEGY?

If such cooperation did typify the radio industry during the period of the introduction of DAB, then that poses a number of questions: did it actually work and did it come to an end? To answer these questions we can examine the latest evidence of DAB's position in the UK radio industry as of 2018, almost two decades since the instigation of a joint effort, and in doing so, it is possible to identify both successes and failings. According to RAJAR figures, by the end of 2017, 62% of the population have a DAB set at home, meaning the platform now accounts for almost 36% of all radio listening hours, thus representing a significant milestone. In fact DAB is considered the main driver of overall digital listening, with the performance of other platforms remaining stable, although mobile phones may represent the main continued threat to the DAB model.[10]

As far as the problem of DAB sets was concerned, sales reached the 5 million mark by 2007 and the 10 million mark only two years later[11] and the 20 million mark in 2014.[12] Despite the increase, the price of DAB sets compared to analogue sets has remained a stumbling block even for those in the commercial sector. While many celebrated the 10 million mark, others lamented the slow uptake with Scott Taunton (Head of Radio, UTV Media) believing the fact that FM radio sales still outnumbered DAB by three to one at this point meant DAB sales were actually 'going into reverse'[13] although sales of analogue sets had also dropped by a significant 18%.[14] Some figures do suggest that DAB may be reaching a point of

[10] RAJAR Q4 2017.
[11] DRDB press release, 30 November 2009.
[12] Digital Radio UK press release, 21 December 2014.
[13] DAB: 10m Sales, But Still to Win us Over. *The Guardian*, 1 December 2009.
[14] OFCOM Digital Radio Report 2012.

acceptable penetration and this was considered to be the case when the main headline for the RAJAR results for the last quarter of 2014 exclaimed: 'Digital Radio overtakes analogue in homes for the first time.'[15] Digital Radio UK was able to celebrate that in-home listening via digital platforms had grown to 46.2% overtaking analogue (45.6%). However, the majority of listening still remained on analogue (56.4%), largely due to the number of people listening in car. The success of DAB is not clear over 20 years after the BBC launched its first trial. While in previous years there was consecutive quarterly growth in digital radio's share of total listening hours this has stabilised since 2013, which may suggest a market penetration of DAB as other digital platforms become more attractive and with sales of DAB sets down by a significant 9.1% over the course of 2013–2014.[16] Despite this, the BBC was able to claim that by 2014 its DAB share of listening 'remained healthy' at 28% compared to commercial radio's 21%[17] although the commercial sector claimed that it was simulcasts of existing analogue BBC services which contributed the most to this figure.[18]

The saviours of the DAB platform may however come from bodies other than the BBC and the commercial sector. As far back as 2006 the electronics retailer, Dixons, was able to announce that it was discontinuing the sale of analogue sets in its stores as it could now offer DAB sets for as little as £29.98 because it now saw the future of radio in the UK as being digital.[19] DAB in-car radio was always seen as a crucial platform and by the first quarter of 2015 Digital Radio UK was able to say there were 450,000 new cars registered with digital radio as standard and, of the top 20 selling car brands in the UK, 5 now offered digital radio as standard in all models (Audi, BMW, Mini, Land Rover and Jaguar) while a further 8 had digital radio as standard in the majority of all models (Ford, Vauxhall, Volkswagen, Nissan, Mercedes, Toyota, Citroen and Skoda).[20] With an estimated 61% of new cars in 2014 having DAB, coupled with the plan to build 182 new digital transmitters by 2016, this meant there was good reason for both Digital Radio UK and the Society of Motor Manufacturers and Traders to

[15] RAJAR Q4 2014.
[16] OFCOM Digital Radio Report 2014.
[17] BBC Platform Report, Q2 2014.
[18] OFCOM Digital Radio Report 2014.
[19] Dixons tunes out of analogue radio sales. Dixons press release, 16 August 2006.
[20] *Radio Today*, 12 April 2015.

join the BBC at a conference at Broadcasting House in February 2015 celebrating this important landmark on the digital path.[21]

A major stumbling block for DAB in the UK has been the issue of digital switchover, 'the point at which all national and large local stations currently broadcasting on both DAB and analogue frequencies will cease to broadcast on analogue.'[22] Committed to the notion of switchover, the government launched the Digital Radio Action Plan in July 2010 to ensure that 'if, or when, the market is ready for a switchover it can be delivered in a way so as to protect the needs of listeners, and results in a radio industry fit for a digital age.'[23] This followed on from the government's 2009 Digital Britain report[24] which confirmed DAB as the future direction of digital radio in the UK. The report also stated that switchover would only take place when digital radio listening figures would reach 50% and when coverage of DAB matched that of FM and it was envisaged these conditions would be met by 2015. The Digital Economy Act 2010 enshrined the need for the UK to prepare for switchover, but despite striving to attain this goal, switchover has proven to be difficult to achieve simply because satisfying the two pre-conditions has proven elusive. As the 2015 deadline has now passed, the future of switchover still remains unclear and a definitive timetable for switchover remains elusive with the government only able to claim it was 'close to its target of getting listeners to switch to digital'[25] without committing to a deadline. Digital radio's relative slow adoption rate compared to digital television, and the subsequent inability to adhere to a date for switchover for radio, highlights the inherent weaknesses within the campaign to promote DAB including a lack of awareness of the medium and scepticism regarding the complete migration of all radio to digital. Lax (2011) perceives the over-optimistic targets as being the result of a digital radio policy that is 'responding rather more to the needs and desires of the industry … than in response to the needs of listeners or to any public consultation or deliberation.'

As well as the government commitment towards DAB, the BBC likewise remains committed as it has done from the very beginning. Having

---

[21] Drive to Digital Conference, 6 February 2015.

[22] Digital Radio Action Plan, 14 February 2013, DCMS.

[23] Digital Radio Action Plan, 8 July 2010, DCMS.

[24] Digital Britain Report, 16 June 2009, DCMS.

[25] Culture Minister Ed Vaizey, interview on BBC Radio 4 Today programme, 12 August 2015.

been involved in developing and then rolling out DAB, the BBC went on to provide much of its infrastructure. This infrastructure took a number of forms. The technical infrastructure involved investment in transmitter coverage which continues, so that in March 2015 the BBC announced it was launching a further 20 national DAB transmitters, increasing coverage from 93% to 97% of the UK by the end of the year.[26] Programming infrastructure reached its pinnacle with the launch of new services in 2002 although commitment to this faltered when a BBC review of Radio 2 and 6 Music recommended the closure of the latter. This was eventually reversed by the BBC Trust following a campaign by listeners.[27] Marketing infrastructure has continued with campaigns for DAB, largely done as part of Digital Radio UK, with the last high profile joint campaign taking place in the run-up to Christmas 2010, although on-air promotion on BBC Radio continues on a daily basis. The BBC still envisages its role very much as leading the drive to digital radio as illustrated by a document produced by BBC Audio & Music in 2010 called Leading into Digital[28] which emphasised the BBC's role in promoting digital radio. However, in defining digital radio, the document describes it as 'audio content and features not provided by FM or AM' which in itself would suggest DAB may have become only one element of a wider definition of digital radio—in other words, a return to platform neutrality. The BBC Radio division's objectives for 2014 centred around making world-class content, transforming offerings to younger audiences, developing a more personal BBC and demonstrating value for money but did not reference DAB at all,[29] although it did feature in the overall corporate objectives with the intention of working to cooperate with government and industry to develop a roadmap for DAB switchover.[30]

For the commercial sector the road to DAB has been more rocky after what one could describe as its apogee around the turn of the millennium. The commercial sector has begun to demonstrate more caution than the BBC, so in 2013 a group of 13 companies called on the government to abandon plans for switchover, claiming the move would jeopardise local

---

[26] BBC press release, 10 March 2015.

[27] Service Review: BBC Radio 2 & BBC 6 Music. BBC Trust, February 2010.

[28] Leading into Digital: What does it mean for BBC Audio & Music? BBC Audio & Music, November 2010.

[29] BBC Radio departmental objectives 2014. BBC Annual Report 2013/2014.

[30] BBC Corporate Objectives 2014. BBC Annual Report 2013/2014.

radio.[31] A problem for the commercial sector has been maintaining a coherent approach to DAB. One can argue that an initial enthusiasm within the sector fragmented, and nowhere was this more evident than in the decision by a number of stations to abandon their place on the DAB multiplexes for purely economic reasons. The most striking example of this came in February 2008 when GCap Media[32] Chief Executive, Fru Hazlitt, announced her intention to close national DAB stations Planet Rock and The Jazz and sell off the company's stake in the national DAB multiplex, claiming that 'DAB with its current cost structure and slow consumer response is not an economically viable platform for the group.'[33] This was a surprising change of strategy for one of the leading commercial players and particularly as it was the successor to GWR which had played such a pivotal role in promoting DAB only a few years previously.[34] Another huge setback within the commercial sector came in October 2008 when Channel 4 announced its decision to withdraw from 4 Digital Group, the consortium which was awarded a licence for the second national commercial DAB radio multiplex in 2007. Channel 4 Chief Executive, Andy Duncan, claimed that his company's DAB ambitions had been hit by a slump in advertising revenue and the need to make savings, he also blamed a lack of enthusiasm among other members of the consortium. Despite proclaiming DAB was the strongest future platform for digital radio, he was clear in ruling out any future Channel 4 foray into digital radio.[35] This decision was a significant low point in DAB history in the UK as it represented a stage where the future of DAB as a medium, supported by all sections of the radio industry, became uncertain after a period of consolidation. It may also represent the point where all-out competition once again replaced cooperation between the BBC and the commercial sector, as Channel 4's decision did not provoke any visible distress within the BBC which had been concerned about the former's aim of competing directly with stations such as BBC Radio 4. However, Ofcom reacted by stating it 'recognises that the economic environment is very

[31] Abandon digital radio switchover plans, stations tell ministers. *The Guardian*, 11 November 2013.
[32] Formed from the merger of the Capital Radio Group and GWR Group in 2005 and subsequently taken over by Global Media in 2008.
[33] *The Guardian*, 11 February 2008.
[34] GCap Media was subsequently taken over by Global Media in 2008 who reversed the strategy.
[35] *The Guardian*, 21 October 2008.

challenging and that all organisations need to make decisions in light of the circumstances they face.'[36] It would be another seven years before a second national DAB multiplex would be awarded.[37]

The future of DAB remains couched in uncertainty and only future historians will be in a position to recount the story of its ultimate success or failure. This study however will be of value as it describes the early period of DAB in the UK. It reveals the reasoning behind the decision by the various parties in the UK radio industry to embrace DAB, it highlights the crucial roles played by the BBC and the commercial sector in pursuing a DAB strategy at both the singular and conjoined levels, and it reveals how this period marked an altering of the established structures within the industry where both the BBC and the commercial sector each played crucial, separate roles in driving that imperative. It is the conjoined role however which represents such a significant shift in policy for both parties. A thrusting policy of cooperation regarding DAB helped launch the technology, and this is something which could not have been achieved by either party on its own. It would be unwise at this stage to attempt to answer any questions regarding the long-term future of DAB, but this study helps us understand its origins and accounts for the status of the industry during this period, thus providing an essential historical appraisal which will be of value to those pursuing later research in this field.

The medium of radio itself continues to occupy a prominent position in a media landscape which has changed greatly over the decades, largely due to its very nature as what McLuhan (2003, 24) described, a 'hot' medium. It is however a history characterised by peaks and troughs and in the second decade of the twenty-first century faces threats from a myriad of sources capable of delivering audio content. Despite this, radio can be admired for its resilience and is still going through a renaissance first identified in the early years of the new millennium (Street 2002, 135).[38] The BBC and commercial radio still have roles to play in this ongoing renaissance, whether that be on DAB or other platforms. Indeed, another example of a collaborative project has been the launch of the Radioplayer[39] console to provide radio content from both bodies on platforms which are

[36] OFCOM press release, 10 October 2008.
[37] Sound Digital was awarded the second national DAB multiplex on 27 March 2015.
[38] Also see Radio Reborn supplement in *The Guardian,* 5 November 2011.
[39] An online application delivering all UK radio stations on a single portal, launched in December 2010.

not traditional radio. Perhaps the next phase of radio's history is characterised by the BBC and commercial radio cooperating to provide multi-platform audio content, but that which we will continue to refer to as radio.

REFERENCES

Born, G. *Uncertain Vision: Birt, Dyke and the Reinvention of the BBC*. London: Secker and Warburg, 2004.

Dimmick, J. *Media Competition and Coexistence: The Theory of the Niche*. Mahwah: L. Erlbaum Associates, 2003.

Hendy, D. "A Political Economy of Radio in the Digital Age." *Journal of Radio and Audio Media* 7, no. 1 (2000).

Howard, Q. (Director of Engineering, GWR, 1982–1987; Chief Executive, Digital One, 1998–2008). Interview with author, 1 May 2004.

Lax, S. "Digital Radio Switchover: The UK Experience." *International Journal of Digital Television* 2, no. 2 (2011).

McLuhan, M. *Understanding Media*. London: Routledge, 2003.

Moudud, J. "The Hidden History of Competition and its Implications" In Moudud, J., Bina, C. and Mason, P. (eds) *Alternative Theories of Competition: Challenges to the Orthodoxy*. Abingdon: Routledge, 2013.

Starkey, G. *"The Quiet Revolution: DAB and the Switchover to Digital Radio in the United Kingdom." Zer* 13, no. 25 (2008).

Street, S. *A Concise History of British Radio 1922–2002*. Tiverton: Kelly Publications, 2002.

Appendix

The Future of ILR

Resolutions passed at the Heathrow Conference
Approved unanimously by a special conference of AIRC member companies at Heathrow on Saturday, 23 June 1984.

1. AIRC is concerned that UK radio developments now being contemplated are examined in the context of all independent radio, and requires the Government, and the Independent Broadcasting Authority, to take full account of the possible effects of any changes or additions to independent radio on the existing ILR system.
2. AIRC requires that any funds drawn from Independent Local Radio by the Independent Broadcasting Authority must not be used for the provision of transmitters, or to meet any other costs, associated with the development of Independent National Radio.
3. AIRC resolves to commission E.I.U (Economist Intelligence Unit) Informatics, as a matter of urgency, to carry out in-depth research into the various levels and consequences of de-regulation.
4. AIRC totally supports the most recent letter from the Chairman of the Independent Broadcasting Authority to the Home Secretary on pirate radio. AIRC resolves that, in the event Government does not take such actions, the members of AIRC will re-consider their own various statutory and royalty payments, currently costing the industry in excess of £13 million a year.

© The Author(s) 2018   217
JP Devlin, *From Analogue to Digital Radio*,
https://doi.org/10.1007/978-3-319-93070-1

5. Recognising the nature of the market place, independent radio companies in the UK should be able to trade with the same degree of freedom as other commercial enterprises, limited only by the explicit requirements of the Broadcasting Act, the Companies Acts and the laws of the land applicable to all businesses and private individuals.
6. AIRC calls upon the Independent Broadcasting Authority to acknowledge the essential differences between radio and television marketing opportunities, and relax the advertising control system which at present prevents ILR companies from seizing specific advertising and sponsorship opportunities.

(Memo sent to all AIRC members, 25 June 1984.)

GLOSSARY

AFN American Forces Network
AIRC Association of Independent Radio Contractors
AM Amplitude Modulation
BBC British Broadcasting Corporation
BSB British Satellite Broadcasting
BTDC Baird Television Development Company
CBC Canadian Broadcasting Corporation
CHR Contemporary Hit Radio
CMA Community Media Association
CRA Community Radio Association
CRCA Commercial Radio Companies Association
DAB Digital Audio Broadcasting
DCMS Department for Culture, Media & Sport
DMB Digital Multimedia Broadcasting
DRDB Digital Radio Development Bureau
DRM Digital Radio Mondiale
DTI Department of Trade and Industry
DTV Digital Television
DVB Digital Video Broadcasting
EBU European Broadcasting Union
FM Frequency Modulation
IBA Independent Broadcasting Authority
IBC International Broadcasting Company
IBU International Broadcasting Union

© The Author(s) 2018
JP Devlin, *From Analogue to Digital Radio*,
https://doi.org/10.1007/978-3-319-93070-1

IFA Internationale Funkausstellung, Berlin
ILR Independent Local Radio
INR Independent National Radio
IRN Independent Radio News
ITA Independent Television Authority
ITV Independent Television
LBC London Broadcasting Company
LRA Local Radio Association
LRWP Local Radio Working Party
Ofcom Office of Communications
PPL Phonographic Performance Limited
PRS Performing Rights Society
PSB Public Service Broadcasting
RA Radio Authority
RAJAR Radio Joint Audience Research
RCA Radio Corporation of America
RDF Radiodiffusion Française
RSL Restricted Service Licence
RTÉ Raidió Teilifís Éireann
VHF Very High Frequency

INDEX[1]

[1]Note: Page numbers followed by "n" refer to notes.

© The Author(s) 2018
JP Devlin, *From Analogue to Digital Radio*,
https://doi.org/10.1007/978-3-319-93070-1